GREEN APPARELS

GREEN APPARELS

Dr. M. Parthiban
Dr. M. R. Srikrishnan
Dr. P. Kandhavadivu

WOODHEAD PUBLISHING INDIA PVT LTD

New Delhi

Published by Woodhead Publishing India Pvt. Ltd.
Woodhead Publishing India Pvt. Ltd.,
303, Vardaan House, 7/28, Ansari Road,
Daryaganj, New Delhi - 110002, India
www.woodheadpublishingindia.com

First published 2018, Woodhead Publishing India Pvt. Ltd.
© Woodhead Publishing India Pvt. Ltd., 2018
Reprint 2020

Woodhead Publishing India Pvt. Ltd. ISBN: 978-93-85059-51-3
Woodhead Publishing India Pvt. Ltd. e-ISBN: 978-93-85059-97-1

Typeset by Allen Smalley, Chennai

Printed and bound in India by Replika Press Pvt. Ltd.

Contents

Preface   vii

1. **Perception of sustainable green consumption practices and it's impact on greener lifestyles**   **1**

   Gopalakrishnan D & Prakash M

2. **Implementation of sustainability initiatives for green production in apparels**   **33**

   SenthilKumar S & Vinayagamurthi P

3. **Eco-hazards in manufacturing of apparels**   **69**

   Manjula S

4. **Dyeing Industry wastewater using reverse osmosis treatment practiced in CETP –Part I**   **83**

   Dr. Rameshkumar M & Dr. Saravanakumar T

5. **Evolution of anti-pollution face mask using three layer composite fabrics**   **107**

   Dr. Srikrishnan M R, Niresh J & Archana N

6. **Eco-management in Apparel Industry**   **147**

   Dr. Parthiban M & Dr. Srikrishnan M R

7. **3P's concepts for Sustainable Development**   **153**

   Dr. Parthiban M & Dr. Kandhavadivu P

8. **Enzymatic approach to sustainable apparel production and green applications**   **161**

   Gopalakrishnan D, Dr. Meenu Srivastava & Apoorva Gupta

9. **Reduction of carbon-footprints in apparel industry**   **181**

   Dr. Parthiban M & Dr. Kandhavadivu P

10. **Eco-testing of apparel products**   **189**

    Amutha K

11. **Environmental Sustainability in Textile Production and** Processing   **221**

    Dr. Senthil Kumar P

Preface

The textile and apparel industry which includes fibres and textiles for technical and apparel production, is one of the world's most important and highly integrated global markets. As the world population increases, consumption of textiles and apparel products also increase. This influences numerous industries within the service and extractive sectors. For example, the textile industry has strong links to energy, water, chemicals, petroleum, agriculture and consumer products as well as to the retail, transport, leisure, sports and high fashion segments of the economy.

All textile production influences the society and the environment. Increasingly, the textile industry has recognized and is taking responsibility for the impacts of their activities in these areas. Voluntary third party certification of sustainability efforts and performance has become an important part of creating and communicating value and transparency by textile firms. There is a clear need for better reporting and certification systems to communicate more clearly how textile factories are addressing the reduction of environmental and social challenges while increasing efficiencies in social, environmental and economic performance.

This book will provide the concepts of developing environment friendly and sustainable clothing for the future. The book focuses on the legal regulations, ecological considerations and different standards recommended by various countries and certifying agencies. It also speaks about the characterization of environment friendly apparel products and the concepts related to the development of earth positive apparels. The book also speaks about the cleaner production technologies for future dwells on novel technological aspects related to wet processing industry.

Dr. M. Parthiban
Dr. M. R. Srikrishnan
Dr. P. Kandhavadivu

Perception of sustainable green consumption practices and its impact on greener lifestyles

D. Gopalakrishnan[1] and M. Prakash[2]

[1] *Department of Fashion Technology, PSG College of Technology,*
Coimbatore – 641 004
[2] *Department of Fashion Technology, Sona College of Technology,*
Salem – 636 005
Email: dgk.psgtech@gmail.com

Abstract: Green fashion provides consumers with healthier and more eco-friendly apparel choices. The fashion supply chain is one of the most polluting industries in the world, being a huge consumer of water, electricity and chemicals, and discharging massive quantities of wastes to land. Going green has seeped into the nation's consumer consciousness. Synthetic fabrics and clothing fibres processed with heavy chemical agents are facing a declining popularity because of the harmful and enduring effects on the planet and people's health. Concerned consumers are also becoming increasingly health conscious and actively support greener lifestyles. A review of the environmental impacts associated with apparel across the entire product life cycle revealed that impacts from the production and processing and apparel consumer use stage dwarf those of transportation and product end-of-life. Global pollution and increased awareness are prompting consumers to seek healthier living choices including clothing in these countries. However, the same has not happened in India. Neither textile manufacturers are doing much for sustaining green fashion, nor the lack of awareness about green fashion among customers driving them to go green not only in their own interest but also in the interest of environment. And while some industries have received more attention than others, research has shown that even for consumers with knowledge of environmental impacts resulting from apparel production and manufacture, purchasing green over conventional apparel has not historically been a concern for many consumers. The business interests of these manufacturers act as deterrents for going green. They said that they need to look at all sides of this issue including the end cost of the process. The areas to look at include not only the cost, but also time and even pollution levels caused by the transition to green. Stakeholder's and customer's pressure on sustainability has pushed companies to transform general environmental sustainability concepts into business practices. This is the reason that green clothing is hitting the runways, not only as a fashion, but as a way of life in many countries around the globe. However, a few contributions have offered a comprehensive analysis of the practices employed in the fashion supply chain to reduce its environmental impact. In this paper, a theoretical perception and impact of sustainable green consumption practices in the greener lifestyle is presented.

Key words: green products, greener lifestyles, green consumption, sustainability

1.1   Introduction

Contrary to the rapid progress made by the textile industry in advanced countries in their efforts to minimise ecological destruction and offer consumers natural products choices, with reduced or eliminated highly toxic ingredients, not much has been done by the textile industry in India. A major driver of human impact on Earth systems is the destruction of biophysical resources, and especially, the Earth's eco-systems. While going green can be a worthy endeavour for any textile industry, there are drawbacks to opt for green fashion and help the green movement in the interest of nation. The environmental impact of a community or of humankind as a whole depends both on population and impact per person, which in turn depends in complex ways on what resources are being used, whether or not those resources are renewable, and the scale of the human activity relative to the carrying capacity of the eco-systems involved. The story of those concerned with textile industry is quite simple. The extent of awareness regarding green fashion among Indian citizens is considered to be quite low. This is the reason that green fashion has not become sustainable in India in spite of the increasing awareness about environmental degradation, environmental pollution and its hazardous effects on the health of the people.

A review was performed across different topics to identify consumer preferences that may influence the implementation of sustainability practices for outdoor apparel brands. Topics reviewed include factors that influence consumers to purchase green apparel and whether the outdoor apparel brand's main consumer base, the outdoor recreationalist, is more likely to be concerned with the environment than the average consumer and how that may affect their preference when purchasing apparel. Lastly, it was explored whether a company or brand's commitment, real or perceived, to environmentally friendly practices or production of environmentally friendly products will help to sell more items.

The birth of the modern green consumerism is usually dated at around the time the newly released Brundtland report (1987) was generating heightened awareness of the global ecological crisis. It defines sustainable development as the development that meets the needs of the present without compromising the ability of future generations to meet their own needs. Erickson (2004) brings out the fact that to minimise environmental impacts by firms requires the blending of good engineering with good economics as well as changing consumer preferences. Rao and Holt (2005) highlights that green supply chain management promotes efficiency and synergy among business partners and their lead corporations, and helps to enhance environmental performance, minimise waste and achieve cost savings. This synergy is expected to enhance the corporate image, competitive advantage and marketing exposure.

De Brito et al. (2008) suggested different practices to pursue environmental sustainability objectives, in terms of both a single company and the whole supply chain. While there already exists a broader consumer market for sustainably sourced goods, as evidenced by the fact that an estimated 85% of US consumers already purchase green products (Grail Research, 2009), the apparel industry has historically not received nearly as much attention as perhaps the food industry where concern has been voiced by consumers regarding herbicide/pesticide usage for grown crops, genetically modified food, and hormone/antibiotic over usage for livestock animals. There also has been a rise in popularity and proliferation of community farmer's markets selling locally grown and organic produce. The lack of attention on the apparel industry however has begun to change. Shrivastava (1995) highlights the importance of organisational sustainability and emphasises that organisational sustainability will require the creation of new organisational processes and systems. This would be geared toward creating inimitable green production systems, first-mover market strategies to capture emerging green markets, ecologically efficient cost structures for long-term profits, a better legal system regarding environmental and product liabilities, and environmental programs for better public relations and community image.

Caniato et al. (2012) presented the results of exploratory case-based research aimed at identifying the drivers that push companies to adopt green practices, the different practices that can be used to improve environmental sustainability. The consumers do not grasp that apparel manufacture and retail can cause significant environmental pollution. These impacts will vary depending on the type of fibre a garment is made from, but they will occur throughout a product's life cycle and can include significant energy use, natural resource depletion, greenhouse gas and other air emissions from processing fossil fuels into synthetic fibres (polyester or nylon); significant water use, toxicity from fertilisers, pesticide and herbicide use related to production of fibre crops (e.g., cotton); and water use, hazardous waste, and toxic effluents from the production stage of apparel that includes chemical usage for pre-treatment, dyes, and finishes; and from product end of use and transport (European Commission, 2013). Chi (2011) analysed the development, achievements, and challenges of sustainability practices in the Chinese textile and apparel industry.

Negative attention has also been given to fast fashion (low-cost clothing that mimics current luxury fashion trends) and how it is predicated upon recent trends quickly running their course and then making way for the next trend (Joy, 2012), with garments usually disposed after being worn ten times or less (Birtwistle and Moore, 2007). For some clothing brands, particularly those in the specialised outdoor gear and apparel sector, supporting environmental causes and espousing environmental activism is not a new idea. In fact, some brands, such as The North Face (Tomlinson, 2011) and Patagonia (Stevenson, 2012) have founders who are noted for their environmental conservation and activist efforts. This attention and subsequent greater demand by consumers for more significant efforts to promote environmentally friendly practices across other industries have not gone unnoticed by the apparel industry. In response, industry groups such as the Sustainable Apparel Coalition (SAC) have been formed to promote, in their own words—An apparel and footwear industry that produces no unnecessary environmental harm and has a positive impact on the people and communities associated with its activities (SAC, 2012). In literature, a few contributions have offered comprehensive and structured analysis of the different practices that have been employed in the textile sectors to reduce their negative impacts.

1.2   Green apparel consumers

Many choices confront a consumer when considering what and how it means to be a green apparel consumer. Brosdahl and Carpenter (2010) stated that whether consumers do not have or could use more information, that education of those consumers appeared to be the key to encouraging more environmentally friendly apparel purchasing. Environmentally friendly apparel purchases can vary and may include purchasing clothing expressly made with minimal impact to the environment; apparel made only from organic materials; or maybe a consumer only looks to purchase quality made products that will last longer than other garments (Chen and Burns, 2006).

Eastwood (2007) highlights while environmental considerations have always been part of the design process in textile and clothing industry, during last 2–3 years environmental awareness continues to become a more pressing imperative; and as a result, designers may look at the source, make-up, and toxicity of raw materials; the energy and resources required in manufacturing the product; and how the product can be recycled or reused at the end of its life—balanced with other product considerations such as quality, price, and functionality. While corporations must bear much of the responsibility for environmental deprivation, ultimately it is consumers who demand goods and thus create host of environmental problems. Therefore even though corporations can have a great impact on the natural environment, the responsibility should not be theirs alone. Hence customers themselves also have to sacrifice their lifestyles in order to achieve ecological sustainability. Ultimately ecological sustainability requires that consumers want a cleaner environment and are willing to pay for it, possibly through premium priced goods, modified individual lifestyles, or even governmental intervention. Until this occurs it will be difficult for corporations alone to lead the ecological sustainability revolution.

Atilgan (2007) emphasises eco-labelling has encouraged the design, production, marketing and usage of the products with minimised environmental effects during its life span, while acquainting the consumer with the environmental effects of the products that they use. Chen (2001) emphasises that the success of green product development and its actual benefit to the environment depend heavily on the joint effort by both the private and public sectors. While the industry should recognise that the call for green products from the public is actually a marketing or economic opportunity rather than an annoying burden or an inevitable threat, the government should create a regulatory environment that is on the one hand benign to green product innovation and on the other hand strict enough to ensure the overall environmental quality. Nimon and Beghin (1999) brings out introduction of an eco-label as an opportunity which allows market to value process attributes and to rewards producers of environmentally friendly attributes.

Zsolnai (2002) argues that it is not possible to achieve ecologically sustainable consumption by large-scale corporations, which aim to maintain their international competitiveness, and to speed economic growth. It can be achieved by small-scale corporations that, rather than trading across the globe, run their own economic affairs in a substantive way to meet or make the most of their requirements from their local resources. The Kim and Damhorst (1998) study also speculated that businesses could even serve to educate consumers further about the environmental benefits of some of their apparel products, and that when they learned about those benefits; some consumers may be more motivated to choose the green alternative. Studies have also been performed to specifically examine influencing factors for consumers when purchasing sustainable or green apparel. One study done in 1998 by Kim and Damhorst explored several themes related to apparel consumption and environmentalism that included exploring consumer's knowledge of environmental issues related to apparel products, concern for the environment, and behaviour that may be brought about because of environmental concern. The study concluded that while there was no strong relation between environmental knowledge and concern for the environment and responsible apparel consumption, it did find that general environmental responsible behaviour was more strongly related to environmentally responsible apparel consumption (Kim and Damhorst, 1998).

A case study performed in Hawaii was reviewed that attempted to profile consumers that would conceivably pay more to purchase organic cotton in place of conventional cotton products (Lin, 2010). The results of this case study showed that the profile of potential organic cotton consumer who might pay higher prices for organic cotton was one who displayed certain pro-environmental attitudes and behaviour that included among others the importance of being environmentally responsible, considered environmental issues when making a purchase, and was involved in environmental organisations (Lin, 2010). Another study performed in 2010 by Brosdahl and Carpenter, did also generally corroborate the above findings, that knowledge alone of environmental impacts from textile and apparel production did not necessarily encourage environmentally friendly consumption of apparel. However, in contrast, this study indicated that environmental concern did positively influence environmentally friendly apparel consumption behaviour and that this concern could serve as a mediator between knowledge and behaviour and ultimately influence and perhaps modify a consumer's purchasing behaviour (Brosdahl and Carpenter, 2010). While another study by Hustvedt (2006) found that consumer likelihood of purchasing an organic cotton T-shirt vs a conventional cotton product decreased as price increased. Additionally, a study performed found that if an eco-friendly product is to be successful in the market, its environmental superiority could not be the only core value added, that it would be successful only if customers perceived the product attributes as superior to other similar product offerings (Meyer, 2001).

A study performed by Gilg, Barr, and Ford in 2005 identified three questions that are needed to identify green purchasers—who buys, what, when, and why? From those questions, three sets of variables were identified as being influential when classifying green consumer's environmental and social values, socio-demographic variables, and psychological factors. And while it was not a surprise, green consumers were found to be individuals who tended to hold more pro-environmental and pro-social values. It was also found that green consumers were mostly liberal and would look to purchase sustainable goods if they perceived that those purchases would have a minimal environmental impact (Gilg, Barr, and Ford, 2005). Christmann (2000) emphasises that before deciding on environmental strategies, firms need to examine their existing resource base and capabilities. Firms should select environmental practices that fit with their existing resources and capabilities. Because complementary assets are created in a firm's business strategy, the starting point for the formulation of an environmental strategy has to be a business strategy and the resources and capabilities it creates. Firms that lack capabilities for process innovation and implementation may be better off implementing environmental strategies later than other firms, so that they can learn from early implementers and imitate successful environmental practices. Even though the above studies indicated that knowledge of environmental impacts of textile manufacturing did not generally influence purchase of environmentally friendly apparel, one common theme from the above reviewed studies was that when consumers were provided with knowledge of the environmental impacts of textile and apparel, this was found to influence their concern for the environment and potentially their consumption behaviour. The above studies have established that the more a person is environmentally conscious and exposed to knowledge regarding environmental impacts from apparel and textile, the more likely that consumer will purchase sustainable apparel. It was also noted in a profile of consumers who did purchase environmentally friendly apparel that some common attributes seemed to be an importance placed on being environmentally responsible and being involved in environmental organisations.

An extensive literature review was carried out to identify the reasons for green consumerism, the role of corporations in achieving ecological sustainability, the ecologically sustainable business practices in textile and clothing industry and the vitality of leverage strength of ecologically sustainable business practices to improve business performance of corporations in textile and clothing industry.

1.3   Life cycle assessment

Some researchers investigated that introducing life cycle assessment (LCA) is an important tool for achieving environment friendly products. Many companies in our country specifically apparel manufacturing industries have recently committed to apply LCA for their manufactured products. Life cycle assessment is such kind of assessment tool helps to inform the consumer about the environmental effects and consequences including eutrophication process, water usage and CO_2 emission during the entire life of the product. This assessment tool helps reduction of water use, CO_2 emission as well as the level of eutrophication process if it followed strictly and in a systematic way. Introducing LCA of the garment products will promote eco-labelling of garment products. In the AUTEX Conference (2011), some researchers found that, there is a strong involvement of individuals in fashion, when the environmental concern has been evolving and the strong need for action in the field of clothing sustainability, especially through the services of designing or redesigning, where the collection of textile waste should be incorporated into all stages of product life cycle to better maximise reuse and recycling. In an article, Anika Kozlowski, Michal Bardecki and Cory Searcy offer a conceptual and analytical framework by conflating life cycle and stakeholder analyses to develop responses for the fashion industry. They exemplify that identification of stakeholders and their interests, responsibilities and accountability that can provide a basis for the development and implementation of appropriate policies and programs to respond to environmental and social concerns within the circumstance of corporate social responsibility (CSR) of the company.

1.4   Green supply chain management practices

Chris, K.Y.Lo, Andy, C.L., Yeung, T.C.E., Cheng (2011) proposed rising environmental concerns from consumers and stakeholder groups, environmental management has become an important responsibility for today's fashion and textiles manufacturers. The production of fashion and textiles related products often requires high levels of energy and water consumption, and emits large quantities of pollutants to the environment. Therefore, the adoption of environmental management systems (EMSs) is important and could have a significant impact on these firms operational performance. This study presents empirical evidence on the performance impact of EMS adoption in the fashion and textiles related industries (FTIs). Handfield et al. (2002) studies prescriptive models for measures of GSCM practices implementation with a focus on GP and GSCM have been developed a decision model to measure environmental practice of suppliers using a multi attribute utility theory approach.

Kainumaa and Tawarab (2006) have proposed the multiple attribute utility theory method for assessing a supply chain including re-use and recycling throughout the life cycle of products and services using the tool of life cycle assessment. Joseph Sarkis, Qinghua Zhu, Kee-hung Lai (2010) proposed green supply chain management (GSCM) has gained increasing attention within both academic and industry. As the literature, finding new directions by critically evaluating the research and identifying future directions becomes important in advancing knowledge for the field.

Qinghua Zhu, Joseph Sarkis, Kee-hung Lai (2007) this study aims to empirically investigate the construct of and the scale for evaluating green supply chain management (GSCM) practices implementation among manufacturers. With data collected from 341 Chinese manufacturers, two measurement models of GSCM practices implementation were tested and compared by confirmatory factor analysis. Our empirical findings suggest that both the first-order and the second-order models for GSCM implementation are reliable and valid. Guo-Ciang Wun, Jyh-Hong Ding, Ping-Shun Chen (2011) have proposed empirical study of Taiwan's textile and apparel manufacturers investigates the relationships between green supply chain management (GSCM) drivers (organisational support, social capital and government involvement) and GSCM practices (green purchasing, cooperation with customers, eco-design and investment recovery). It also studies moderating effects by institutional market, regulatory and competitive pressures, the results of this research show that (1) except for investment recovery, the other three GSCM practices are positively affected by GSCM drivers; (2) investment recovery is positively affected only by organizational support; (3) market pressure has no moderating effects on most of the relationships between GSCM drivers and GSCM practices; (4) regulatory pressure has positive moderating effects on most of the relationships between GSCM drivers and GSCM practices; and (5) competitive pressure has negative moderating effects on most of the relationships between GSCM drivers and GSCM practices. Qinghua Zhu, Joseph Sarkis (2004) gives green supply chain management (GSCM) for developing country, and examine the relationships between GSCM practice and environmental and economic performance. Using moderated hierarchical regression analysis, evaluate the general relationships between specific GSCM practices and performance. Two primary types of management operations philosophies, quality management and just-in-time (or lean) manufacturing principles, influence the relationship between GSCM practices and performance. Significant findings were determined for a number of relationships. Managerial implications are also identified.

1.5   Textile and clothing industries

Food, shelter, and clothing are the basic needs of everyone. All clothing is made from textiles and shelters are made more comfortable and attractive by the use of textiles. Everyone is surrounded by textiles from birth to death. We walk on and wear textile products; we sit on fabric-covered chairs and sofas; we sleep on and under fabrics; textiles dry us or keep us dry; they keep us warm and protect us from the sun, fire and infection. Clothing and household textiles are aesthetically pleasing and vary in colour, design, and texture. They are available in a variety of price ranges.

1.6   Changes in textiles

This text was written for consumers—not average consumers but educated consumers who, when they purchase textiles items, want to know what to expect in fabric performance and why fabrics perform as they do. Textiles are always changing. They change as fashion changes and to meet the need of changing lifestyles of people. New development in production processes also cause changes in textiles, as do government standards for safety, environmental quality and energy conservation. These changes are discussed but the bulk of the text is devoted to basic information about apparel and household textiles with an emphasis on fibres, yarns, fabric construction, and finishes. All of these elements are interdependent and contribute to the beauty and texture, the durability and serviceability, and the comfort of fabrics and garments should be re-conditioned. Verma (2002) did a comprehensive study with objective to evaluate the export competitiveness of Indian textile and clothing sector. Because Indian textile and clothing sector is predominantly cotton based, the study is focused on cotton textile and clothing and look at the entire value chain from fibre to garment and retail distribution. The scope of study covers the products in Indian export basket which have shown a promising growth in value. The study concludes that Indian exports to US and EU are export competitive as a whole. Sector wise analysis of export performance of Indian textile and clothing sectors to US and EU reveal that so far apparel or clothing and made-up is concerned; quota is the major constraint in the growth, while it is not true in case of yarn exports. Indian textile and clothing sector has tremendous potential and only a portion of which is explored till now and this shortcoming is due to policy constraints.

Simpson and Shetty (2001) did a vast study on India's textile industry. The purpose of study is to analyse India's textile and apparel industry, its structural problems, market access barriers, and measurements taken by government of India to enhance the industry's competitiveness in the post— Multi-fibre Agreement (MFA) era. The study also assesses India's textile and apparel market potential and trade and investment opportunities for US firms as India steps into a more free and transparent trade regime. For the purpose of study, exploratory study is done in which in-depth interviews are done with various government officials in Textile Export Promotion Council, Ministry of Textile, Cotton Council of India, Apparel Export Promotion Council (AEPC), Federation of Karnataka Chamber of Commerce and Industry, Handloom Export Promotion Council, Madras Chamber of Commerce and Industry, The South India Textile and Research Association, and almost all top executives of India's large textile mills.

Chandra (1998) in his research wrote on challenges ahead of Indian textile and clothing industry in post quota regime. It put special emphasis on production capabilities and efficiencies as most essential elements to fight global competition. It suggests various strategic decisions Indian textile manufacturers have to make to survive the competitiveness in post quota regime. Vera (2001) has emphasised on the impact on the Indian textile and clothing industry after quota elimination. It says that Indian textile and clothing exporters have to bring in necessary changes in their methods of production, management style, capacities, marketing skills and productivity level in order to remain competitive in international market. Also it put special emphasis on the size of Indian textile units when compared to its counterpart in China. Meenakshi (2003) did a comprehensive study on the opportunities that would be provided by WTO (World Trade Organization) to Indian textile industry. This paper gives a lot of emphasis on new capacity installation to take the benefits to the fullest extent in India has to be a true gainer in competition to other nations. Since India's own consumption per capita is also on the rise with the rise of income and consumption habits, the profit margins available to Indian textile and clothing producers will be more. But in export market, the prices will be driven by international factors and profits will be under pressure. So the exporters might have to go for strategy of partial exports and partial domestic sale.

Thomas (2005) in his article wrote on why in the competitive scenario wholesalers like Nike are shy from keeping long inventories and stocks. So pressure is on garment companies to deliver the goods in time. India has bottleneck in infrastructure, which hinders the time receipt of raw material and delivery of finished goods. This would cause rapid freight and would squeeze the margins. Government has to invest heavily in infrastructure to keep the pace of growth of garment industry intact and take the benefits to fullest extent. Pandey (2003) in his article expected that Indian textile exporter would be benefited with quota elimination. It discusses on various sectors of textile and clothing. Also he expects that hosiery industry will be one of the gainer and small scale exporters will be more competitive due to small size and controlled cost and lower overheads. Uraiwan (2004) had worked extensively on the knitwear/hosiery products development process to understand the complexities underlying in it; because a well-defined development process assist the organisation to determine its future direction, plan for rapid changes, create new product line with profits and plan for technology adaptation and implementation. The goal of this research was to propose an optimal product development process for a knitwear/hosiery company by examining the process used by major US Sweater Company and comparing its process to established processes. Bedi (2009) in his article had prepared detailed report on Indian textile industry covering various sector of textile industry. This is one of the most comprehensive reports coveting all aspects of textile industry, performance and hindrances in the growth of it. Vivek (2004) in his article had said that JC Penny a leading retail chain of US looks India for sourcing its garments in woven and hosiery. He is of opinion that India will be fulfilling its major need of hosiery and woven garments in cotton while China will be good for synthetic fabrics and its garments.

Chugan (2005) emphasised that Indian textile industry has to change to be more competitive in the long run. This paper emphasis that merely cost competence is not enough to maintain the lead while Indian companied has to have a global competitive view. Trivedi (2005) in his article concluded that the textile is one sector where India has high ambitions and can achieve robust growth through moderate human skills. India has skilled labour and does better in this sector as compared to others. This will also increase the employment and the social structure will be better off. Chaudhry (2006) did a very comprehensive study on the productivity of Indian textile sector and various related sectors. Very technical formulas are used to analyse the competitiveness of Indian textile industry. Chugan (2006) in his article discussed in detail the opportunities available to various sectors of Indian textiles in the post quota era. Also, it emphasises the weaker link, competition from china and the schemes run by government to support Indian textile industry. Kumar (2006) did study of various sectors of Indian and Chinese textiles. This paper concludes and highlights the various areas where India has efficiency over china and how India should more capitalise on it. Also it gives equally weightage to Chinese advantages and how India can win over its weaker areas to be more competitive in long run. Texprocil (2007) in that article concluded that if India has to keep maintaining its edge in hosiery and garment sector, it has to keep in control through various measures. The various measures indicated are raw material, methodology, labour wages, power cost and utilities that need to be kept in check to keep the cost lower. This paper presents a comparative study of Indian textile industry with other nations like China, Bangladesh, Vietnam, Egypt and Pakistan and elaborates the competitiveness of Indian textile and various sectors in textiles. It also puts lots of emphasis on the areas where India in losing its edge and has to keep a close monitoring on it to remain competitive. It concludes that Vietnam and Egypt are coming up fast and can prove to be tough competitor in near future due to high productivity and low steam cost.

1.7   Law/act for textile industries

To make textile selection a bit easier consumers, textile producers and their associations have set standards and established quality control programs for many textile products. The federal government has passed laws to protect consumers from unfair trade practices, namely, The Wool Products Labelling Act, The Fur Product Labelling Act, The Textile Fibre Products Identification Act, and The Flammable Fabrics Act. The first three laws are—truth in fabrics legislation and to be beneficial knowledge on the part of the consumer about fibres and furs is required. The fibres most commonly used were wool, flax, cotton, and silk.

1.7.1   Man-made textile products

Silk has always been a highly prized fibre because of the smooth, lustrous, soft fabrics made from it; it has always been expensive and comparatively scarce. It was, therefore, logical for man to try to duplicate silk. Rayon was the first man-made fibre. Rayon was produced in filament until the early 1930s when an enterprising textile worker discovered that the broken and wasted rayon filaments could be used as staple fibre. Acetate and nylon were also introduced as filaments to be used as silk like fibres.

1.7.2   Consumption of bio-resources

The scale of the human activity relative to the carrying capacity of the eco-systems involved. Careful resource management can be applied at many scales, from economic sectors like agriculture, manufacturing and industry, to work organisations, the consumption patterns of households and individuals and to the resource demands of individual goods and services. One of the initial attempts to express human impact mathematically was developed in the 1970s and is called the IPAT formula. This formulation attempts to explain human consumption in terms of three components—population numbers, levels of consumption (which it terms "affluence", although the usage is different), and impact per unit of resource use (which is termed "technology", because this impact depends on the technology used). The equation is expressed as

$$I = P \times A \times T$$

where
I = environmental impact, P = population, A = affluence, T = technology.

The research in 2005—philosophy and ethical consumption—authors initiated some basic philosophical approaches that are practical in understanding and evaluating ethical consumption issues and ethical consumer behaviour. They have argued that such formalistic philosophical positions can be too demanding and abstract for application in everyday consumption. In another research paper, authors looked into consumers' perspectives on sustainable clothing consumption and examine some specific ways in which information could pressure retailer's policies. In the year 2006, some scholars introduced better understanding of ethical fashion consumption; even though consumers demand more ethical responsibility from companies. In an article, researcher shows the structure of sustainable fashion supply chain including eco-material preparation, sustainable manufacturing, green distribution, green retailing, and ethical consumers based on the extant literature. The fashion industry today is a global industry and has a huge effect on our environment as well as on people. It is dominated by fast fashion and just-in-time production that has led to increased seasons and mini collections in season, which generate new low price items in store every week and even every day.

1.8   Environmental concern and ethical issues

Many growing factors considered which are distinguished ethical from traditional fashion including use of sweatshop-free labour, energy-efficient processes, alternative energy and low impact dyes in manufacturing. Nevertheless, fashion consumers now-a-days are trying to choose an ethical wardrobe to pick up eco-friendly garment or fabrics. There are three criteria for selecting eco-friendly fabrics as

1. The use of fewer toxic chemicals.
2. The use of less land and water.
3. The reduction of greenhouse gases.

In fact, some of the fabrics may perform better than others based on the above criteria. It may in more cases, one fabric is more preferable according to one of the criteria but less preferable according to another, making for complicating choice in fabric qualities, cost, labour conditions or carbon footprint of product transportation. Now, many cotton firms or industries have a vibrant campaign promoting their products as sustainable pointing that due to new technologies and farming methods. The industry has greatly reduced its use of energy, water and toxic chemicals. But very few farming has the significant success of reducing soil erosion, improved irrigation methods to reduce water use, improve methods of pest management, have reduced pesticides and most significantly, the use of genetically modified (GMO) cotton has reduced the use of land and toxic chemicals. A general statistics has shown that, growing enough cotton to make a single cotton T-shirt a third of a pound of toxic chemical (including pesticides, fertiliser and defoliation chemicals).

1.9   Green consumption practices

Carmen Tanner and Sybille Wolfing Kast (2003), in their article have attempted to know the barriers faced by consumers in purchase of green foods and gaining knowledge about green purchases of Swiss consumers.

They found that green purchases are not significantly related to moral thinking, monetary barriers or the socio-economic characteristics of consumers. Irene Tilikidou (2007), in the article has discussed about the pro-environmental purchasing behaviour of Greeks. The study revealed that majority of respondents adopted energy and waste conservation, reduction of overall consumption and they avoid products which are genetically modified. They prefer environment friendly products to others when there is not much difference in price. Alice Gronhoj and Folke Olander (2007) have discussed about the difference in male and female pertaining to consumer behaviour towards environmentally related family consumption. The sample included 30 couples with children. It was found that there was not much difference in the responses given. The researcher has suggested that there should be keen research attention given to environment oriented consumer practices which are adopted and transmitted among family members.

Nik Ramli Nik Abdul Rashid (2009), in the article has researched upon the awareness of eco-label in Malaysia. He has introduced eco-label as a separate variable. The study was made among the employees in organisations which adopted environment management system. The study revealed there will be a positive reaction towards eco-label when consumers are made aware of environment related issues. Iain R. Black and Helen Cherrier (2010), has discussed about the anti-consumption practices as a part of sustainable lifestyle. They have also suggested that the marketers are to position sustainable practices such as independence, quality or value for money. Jūratė Banytė, Lina Brazionienė, Agnė Gadeikienė (2010), in their article has suggested that a green consumer is characterised as an educated woman who belongs to the age group from 30 to 44 and receives higher than average monthly income. Deborah J.C. Brosdahl and Jason M. Carpenter (2010) have discussed about environment friendly consumption behaviour in textiles and apparel. It was found the awareness about production process among consumers of textiles and apparels leads to environment friendly consumption behaviour.

Shilpi Sharma and Maneesha Shukul (2012) according to their study conducted on 75 women consumers about their eco-friendly behaviour while purchasing packaged goods from Vadodara, it was found that respondents had a moderate extent of eco-friendly buying behaviour and the reasons behind the purchase of selected goods in different packaging were economy and convenience with very few reflecting environmental concern. Sung Hee Park, Kyung Wha Oh and Youn Kyu Na (2013), in their article have discussed about consumer attitude towards environment friendly products specifically artificial leather products. The survey revealed factors such as public participation, recycling and resource conservation as having an influence on environment friendly attitudes. Muruganandam D and Gopalakrishnan

S (2013), in their article have brought out the buying behaviour of green products among consumers. They have studied the impact of advertisement on green product purchase behaviour.

1.10   Sustainable organic clothing

The global textile industry is reported to be the most unsustainable; it produces large amounts of waste and discharges into the industry. For example, cotton crops alone consume 24% of the total pesticide used globally (Quinn, 2010). This causes a serious impact on the environment. In relation to fashion, the forecasting process and the continual updating of trends, some of the contributing factors are an increase in product lines and subsequently causes drain on resources; including dyes and chemicals which are discharged into the environment. Subsequently, this affects quality and due to the fast fashion trend, garments lose their face value and usable life much earlier than expected. In light of these detrimental factors, organic natural fibres had shown potential for the women's wear market. Yarn Expo 2011, Beijing, an exhibition which focused on sustainable materials, reported that there was a significant demand for sustainable organic fibres such as cotton, bamboo, flax, and ramie (Ecotextile, 2011). However, performance of garments made of new organic blends such as bamboo and soya, are yet to be documented. In this context, the current paper proposes to investigate the benefits of clothing made of organic natural fibres which offer product serviceability qualities such as durability, longevity and aesthetic appeal.

The popularity of organic clothing has increased in the last decade or so, particularly due to continuous change in consumer lifestyle and increasing awareness of organic produce launched by many major retailers and supermarkets. Consumers are willing to pay more if organic clothing offered durability and longevity. Longevity of clothing also has a major influential impact on clothing retailers who frequently supply fast fashion garments which positively increases their sales. However, according to the Carbon Trust, UK (2011) long lived clothes offer an opportunity to reduce emissions associated with clothing over a long term basis. In its report, Carbon Trust stated that a shirt lasting 6 months will need to be replaced twice a year; doubling the embodied emissions over the year compared to a longer lived shirt that lasts for 50 cycles. The report also stated that by reducing the useful life of clothing from 1 year to 1 month increases the emissions over the year by 550%. Organic fibres that are produced under a natural process without the use of any artificial methods or chemicals are getting more and more attention especially in the clothing and fashion industry. A report from the ICAC Recorder (2009) highlighted that India, China and Syria were the highest producers of organic cotton. In addition, the use of organic cotton by major brands such as Marks and Spencer and Timberland had increased in the recent past.

The Soil Association reported that retail sales using organic clothing and textiles reached £100 million, which was a 40% increase in sales and is a ten-fold increase since 2002 (The Soil Association, 2011). The Soil Association is a charity organisation that promotes and develops sustainable approaches to food, farming and other products and is one of the oldest organisations which certify 80% of produce to Global Organic Textile Standards (GOTS) an international standard from field to final finished product. Sustainable products using environment friendly farming methods preserve an ecological balance thus avoiding depletion of natural resources are gaining momentum in fashionable clothing particularly in the UK. In today's age, garments are produced too quickly that are easy to care for and maintain and equally lose their face value. Usually, these garments are trend driven and disposed quickly by consumers, mainly for the loss of aesthetic appeal rather than its functional purpose. Hence, consumers tend to dispose clothing within a short span of time. In the US, according to a report on Municipal Solid Waste, the longevity of textiles was an issue for waste disposal. Clothing was classified as non-durable and it generally lasted less than 3 years (US Environmental Protection Agency, 2010). DEFRA (2004) in its report on fibre use in clothing stated that 58% of fibres used globally in clothing were synthetic fibres. Among synthetic fibres, polyester accounted for 77% followed by other fibres such as nylon (9%), acrylic (6%) and 7% cellulose. In terms of natural fibres, 28% of cotton was used in clothing globally. Hence cotton and polyester were fibre dominants globally. Other natural fibres which are used in clothing were, 10% jute, 4% wool, 4% flax and 4% other fibres (silk, kapok, hemp). Recently a number of natural fibres such as bamboo, soy, corn, banana and pineapple fibre blends were introduced.

Mukhopadhay et al. (2009), reported that natural fibres such as banana fibres, (a bast fibre obtained from the pseudo stem of the banana plant) had good mechanical properties. The breaking strength of fibres for lower linear density fibres (0.25–0.98 tex) was better than higher diameter fibres (3.0–7.0 tex), the authors noted that fine fibres had better structure and further added that the mechanical properties of fibres were affected by a natural variation in plant and processing stage. Yueping et al. (2010) based on their extensive analysis on fibre morphology reported that bamboo fibre (degree of crystallinity) was similar to that of jute. The cellulose matter of bamboo fibre consisted of 73% cellulose, 10% lignin and 12% hemicellulose. These fibres were popular in summer/spring clothing ranges. In addition, a wide range of animal fibres such as angora, alpaca, llama, vicuna, cashmere and mohair were also popular in winter clothing where thermal comfort was required. The Ecologist, 2011, the Director of Young British Designers Debra Hepburn suggested that a new evolution had begun with ethical and moral values, as to how industry conducts itself towards the environment and people; with transparent practices that produces brands that are genuinely honest and ethically sound within their values. She also added that it all started with recycled materials, however, designers are now increasingly focused towards sharing ideas and practice in sustainable production.

Ninimaki (2010) also highlighted the importance of performance, durability and longevity of products made from natural fibres. The term green fashion is also gaining momentum in the high street, mainly focusing on the industry to clean up its act in relation to conditions of manufacturing, fair labour policies and environmentally responsible actions. Hence, green fashion is placing an emphasis on CSR for the industry to adopt. Ticolau (2010) and Gam (2011) stated that eco-conscious consumers evaluated the cost of any garment against its durability and performance and therefore expected garments produced from natural fibres to be superior to man-made garments. Eco-conscious consumers also rely on fabric performance and preferred product serviceability such as comfort and appeal which most eco-friendly garments do not provide (Gam, 2010).

1.11   Eco-friendly design choices for consumer

Eco-friendly apparel designers put the needs of the clothing consumer first in the product criteria, and then find ways to meet those needs that reduce net environmental impact with special attention to how the consumer will actually use the product. Eco-friendly clothing supplies similar, or hopefully superior, quality and performance compared to conventional clothing in terms of fit, durability, and style, which, according to Kardash in Peattie (2001), ensures that any consumer, green and conventional, will be interested in purchasing them. The eco-friendly objective of better meeting consumers clothing needs encouragement to consumers to actually purchase less clothing overall because improved eco-friendly clothing will fit them and last longer. In order to accomplish this, consumers must realise the personal benefit of investing in fewer, high-quality clothing items that will better meet their needs rather than wasting money on cheap clothes that do not fit or last, which in turn benefits the environment through reduced overall resource consumption. The Green Behaviour: How will the consumer use it? Instead of considering the object, the garment, as the focus of your design activity, visualise someone wearing and moving and enjoying her or his life in the garment that you design. Move your focus away from the object to the person. In that way, the shift toward sustainability can begin.

Many accepted principles of eco-friendly design can have a dramatically negative impact in the hands of consumers (Hethorn and Ulasewicz, in press). For instance, eco-efficient light bulbs require less energy to use yet people tend to keep them on longer using more energy than they previously did (van de Velden, 2003). Eco-friendly design goals advocate shifting from selling products to selling services, yet in cases where a product is necessary for providing that service, this actually increases product consumption, such as cell phones that are constantly replaced with the latest upgraded models. In general, most efficiently-produced products result in cheaper prices because they require less energy and materials to make, but leaner products tend to become throwaway goods and, for this reason, to proliferate (Manzini, 2001, p. 4). This is known as the rebound effect (Lilley et al., 2005; Manzini, 2001; van de Velden, 2003) when a consumer uses or often misuses a product or service in a way that was unintended by the designer resulting in an unexpected, often negative, effect on the environment and/or society. Many apparel designers and companies are still unaware of how people use and interact with their clothing and how those behaviours can have an enormous effect on the environment. But some of the more 49 environmentally responsible apparel designers and companies do consider some aspects of the consumer's role. They recognise the importance of reducing the energy consumers use to clean and care for their clothing (Fletcher and Goggin, 2001), as well as offering high-quality, appealing clothing designs that consumers will want to purchase and wear. In order to prevent the —rebound effect, designers must consider what Jelsma and Knot (2002) term—use(r) logic‖: To increase chances for intended outcomes, such normative (re-) design efforts have to start with careful mapping of the interactions between users and their material surroundings in the reference situation, especially with respect to underlying values and logic.

A design logic that aims to inscribe eco-efficiency in products and services while being insensitive to use(r) logic makes little chance to enrol users in new ways. Designers must be aware of and, better yet, know how to influence use(r) logic if they want to achieve sustainable clothing consumption. For example, if eco-friendly clothing were designed to offer better, adjustable fit, the desired outcome would be that consumers would buy less clothing over time. Yet this may go against use(r) logic. As previously mentioned, poor fit often limits what people will buy, so if it were easier for people to find clothes that fit them, they may buy more clothes than they previously did. In the UK during the last 10 years, new clothing sales volume has increased 60% (Morley et al., 2006) while the price of women's clothing has decreased 34% (Beckett, 2006), suggesting that consumers are buying more cheap clothing. Very rarely, and never as continuously, have prices gone down so dramatically as consumer demand has increased, yet this trend is expected to continue indefinitely (Beckett, 2006). Whether this dramatic increase in clothing consumption is mainly due to the 50 lower prices that allow consumers to purchase more for the same amount or to the lower quality of cheap clothing that need to be replaced more often is unclear, but it is probably a combination with both reasons reinforcing each other. The Guardian, a UK newspaper, interviewed a woman leaving a value retail shop who had purchased the same cheap handbag in nine different colours; when asked why, the woman replied—You never know when a bag is going to come in handy when they're £3 a time (Beckett, 2006), demonstrating a more for less mentality. Yet a UK consumer pool showed that over 60% of participants noticed that in the last 3 years, clothing lifetimes had decreased and clothing was becoming lighter, showing that current clothing has a deceased usefulness and lower quality (Morley et al., 2006). Ironically, in making products more efficiently and from fewer resources, steps that should have helped the environment only made products cheaper so that consumers could buy even more. Consumers either bought more of the same products or spent the extra money on other, more environmentally unfriendly expenditures, such as long-distance travel, that resulted in no net environmental improvement (van de Velden, 2003). In order to achieve social and environmental fairness, Western countries need to decrease overall consumption levels and share environmental resources with the rest of the world. Some may argue that it is impossible to change human behaviour, but appealing to some deeper value and logic can influence people to change their ways in order to better reflect those inner values. Reducing consumption in the apparel industry may be helped by higher prices as a result of making better fitting, better quality products. Better fitting more consumers would likely require a greater number of sizes, which would mean lower production levels per size or even custom production. Better quality may use more resources and labour but still could be environmentally and socially responsible. Paying a fair price for environmental resources and sewing labour will most likely require a 51 higher price. Phil Patterson, representative of the socially responsible UK retailer Marks and Spencer, argued that the future of retail should be higher prices for better products, since lower prices did not help anyone as greater volumes of cheaper products only made consumers buy and waste more (personal communication, November 16, 2006).

1.12   Green purchase perception matrix

Peattie (2001) attempted to explain why so many green products failed to achieve consumer support by analysing consumer reactions to all of the attributes of the products. Peattie theorised four possible categories of green purchases to be functions of the consumer's perceived degree of (1) confidence in the product (brand name, environmental effectiveness, etc.) and (2) compromise (product quality, product cost, search cost, information cost,

cost of change, etc.). Confidence in green purchases has been a problem for consumers who mistrust marketers' claims and motivations (Peattie, 2001; Meyer, 2001), and many consumers have been unwilling to compromise product quality for reduced environmental impact. The best —win–win green purchase scenario occurs when the consumer has a low degree of compromise in terms of cost and quality and high degree of confidence in the green product's environmental benefits. For example, energy efficient home design saves the consumer money by reducing or eliminating utility expenses, which translates into reduced greenhouse gas emissions.

Other researchers support Peattie's claim that green products must perform at least similar to their conventional counterparts, if not better, if they are to be embraced by consumers, regardless of a person's environmental concern. A study by Dickson (2001) showed that consumers were more likely to purchase a product based on its traditional product traits, i.e. cost and quality, than on their personal attitudes toward social responsibility. The research findings of green product researcher Robin Roy (1997) further supports Peattie's claims: Any successful greener product must balance environmental performance against the many other design attributes—performance, reliability, appearance, etc.— wanted by the market. Products had to be competitive in terms of performance, quality, and value for money before environmental factors entered the list of consumer requirements. Therefore, when marketing green products, the needs and desires of consumers must be addressed first and foremost while the product attributes that benefit the environment and other social issues should be presented as an added plus (Donaldson, 2005; Ottman, 1999). Manzini (1992) considers green consumerism to be part of—a demand for a new quality. Today, more green companies focus on offering superior performance, competitive prices, and transparency about environmental claims (Donaldson, 2005). By striving to achieve superior performance and consumer appeal, eco-friendly design can set the standard for production and consumption rather than be viewed as a mere alternative to conventional design.

1.13   Consumer perception about eco-friendly apparels

In order to analyse the consumer's perception about ecological friendly consumer perception it is important to understand the fundamental definitions related to the concept of what environmental friendly or socially responsible industry means. In today's world of apparel and fashion the consumers are constantly coming across a number of definitions and terminologies that define the eco-friendly or ethically responsible clothing and fashion industry. Sometimes so many different terms moving in the fashion industry also makes hard for the consumers to make the eco-friendly purchases hence it become very hard for them to differentiate between two products or decide what actually eco-friendly is. Therefore for true examination of the analysis of consumer perception about eco-friendly apparel and textile, it is very important to first look in to the definitions of the few important terms (Thomas S, 2008).

1.14   Eco-fashion

The term eco-friendly is not that simple however. Beard (2008) says retail corporations have to ensure that eco-friendly claimed business has to be taken cared of throughout in whole supply chain from top to bottom. For example the raw materials are not exploiting the resources, harmful chemicals are avoided, labour rights have not been exploited and fair trade policies are considered. Also the transportation of goods is not contributing a negative impact to environment of the society. Eco-friendly is related to modern day fashion and textiles. Thomas S (2008) states eco-friendly apparel marketing and ads can be seen and heard in print and television media, articles, fashion magazines and retailing industry campaigns. The history of the term can be located in the early 90s and it used for the apparel and garments or other fashion or textile related merchandise that is not leaving any negative impact on environment or society. Workers are paid with minimum state declared wage rate and they are also provided with good living if needed. Fair-trade practices are also related to the use of the fibre also regarding the use of chemicals, pesticides and other harmful substances are avoided (Thomas S, 2008). Since almost 80% of the production of apparel and textiles is going on in the developing countries therefore it is very pivotal to ensure for the retailers that the working conditions of the people associated with the industry are good.

1.15   Sustainability

In environmental science, (Peter Berck, 2015), it defined as the ability of the natural systems to remain varied and prolific. For example natural forests and wet lands that are still living are the examples of sustainable natural systems. Conventionally sustainability is the durability of the natural and environmental systems.

1.16   Organic

The term "organic" is related to the production of fibre and textiles in natural way. It means the production of apparel and fibre such as cotton is done with minimum 70% organic or natural fibres. The biggest and most authentic certifying organisation for the organic products are global standards organic which assigns a GOTS (Global Organic Textile Standard) to the processes of textile which are made up of natural fibre while maintaining high level purity throughout the supply chain of the textile. It requires the producers to comply with the social standards criteria, as seen in Figure 1.5, GOTS logo awarded to the certified companies. The producers need to ensure that chemicals and dyes used in the process are not or least harmful to the environment and that the manufacturing site, socially the wet treatment plants must have water treatment plant to treat the industrial waste affluent which is potentially harmful for the environment if released in the natural water bodies (GOTS, updated 2012).

1.17   Ethical fashion

Term "ethical" is also used as a replacement for eco sometimes but ethical is politically more authentic. The use of term is related to morally responsible business practices. In terms of textile and apparel fashion industry, it represents the positive impact of fashion production, designing, consumption, retailing and disposal to the society (Thomas S, 2008). The ultimate beneficiary of ethically responsible practices can be human beings, environment, animals and consumers. For example consideration of natural resource depletion, human working condition, animal skin and leather use in order to manufacture fashion merchandise.

1.18   Environmental friendly

Thomas S (2008) referred to mostly natural fibres because of their ability for biodegradability and endurance with easy to wash and care attribute and better prospects for recycling and reuse.

1.19   Green textiles

Green textiles are referred to the apparel and textile products that have minimum impact on environment. Especially in last decade it has seen that consumers now have more eco-friendly textile, apparel and fabrics available in the market. Especially bay clothing, drapery, upholstery and other fabrics used in apparel and fashion merchandise manufacturing. In addition to that manufacturers are more and more getting concerned to produce merchandise that are free of toxic dye stuffs and other harmful chemicals like heavy metals and pesticides and they ensure that the apparel production is not having any negative impacts on the natural resources like water, land and air quality and are also recyclable (Ballard, 2014).

1.20   Green washing

As the concept of green products, eco-friendly products and consumption is looming over the apparel and textile industry, so does the prevailing encounters of false claims and misleading information from the manufacturers are disclosing on the law makers. According to Kewalramani and Sobelsohn (2012), the term "green washing" is used for the companies that make claims and advertising and promotional campaigns on the basis of going green, however, actually they are not implementing as much ethically responsible business practices as they are claiming to do so. Because eco-friendly, green and organic terminologies are becoming so popular it has become very easy for the companies to utilise the opportunity of making business through using these professional jargons.

1.21   Recycled apparel

In apparel industry recycled apparels are those which are reused after used and disposed of already. They are either sold at used clothing shops, charity shops or are recycled to make new garments and resell in to the market. Recycling and reuse of apparel involves collection of the used garments. Gail Myers, (2014) reminded the remanufacturing and reprocessing of the apparel to give them new life and then again resold in to the market. Due to the fast fashion production and increasing rate of consumption textile and apparel industry is producing huge volumes which can be recycled and reused. Textile material can be degraded to use in the pulp industry, they can be used to stuff toys and other textile home fashion products and they can also be used to decompose and again construct in to new garments to sell in the market.

1.22   Sustainability in fashion

In 2011, Kathryn Reiley and Marilyn DeLong investigated that, sustainability in fashion is going to require a radical change in the practices of all together—designers, manufacturers, marketers and consumers. But customers especially need a spirit for sustainable fashion practice. However, in their research they wanted to examine fashion practices related to consumer's passion for a unique exterior and sources of clothing attainment. For the study, they have taken sample from 97 females of the University students—Midwestern University in the USA and has completed the Desire for Unique Consumer Products (DUCP) Scale developed by Lynn and Harris. The result of the research is if we inspired, such individuals could become a guilt-free model for sustainable practices in the future. Based on a customer study about the intent to contribute in different programs, such as 1:1 funding, ecological design sourcing, an improvement style contest, redesign consulting and an eco-fashion gallery, eco-friendly design sourcing programs have the highest intent to participate of all the studied sustainable social programs. The truthfulness of the design of a fashion garment gives it worth, makes it more attractive, and distinguishes it from the everyday of conventional fashion that has been manufactured off shore. In other study, it was observed that, fashion companies should strongly consider the product development process and extend stewardship across the multiple life cycles of products. In another scholarly works, Maarit Aakko and Ritva Koskennurmi-Sivonen revealed a theoretical model, which illustrates together the elements of sustainability and fashion design. The aim of that model was to serve fashion designers, who wish to take sustainability into consideration. Another research paper, researcher illustrated that, the consumers face bewilderment of information and knowledge, when striving for ecologically sustainable lifestyles and consumption practices; consequently how in the midst of these discursive struggles consumers simultaneously mobilise alternative strategies for sustainable consumption.

At the present phase of industrialisation, manufacturing environment friendly products in a sustainable way is the most important and emerging issue. The main focus comprises not only to the products quality sustainability but also it focuses on the manufacture processes including raw material resources from cradle to grave. Now-a-days, many companies and organisations focus on the environment friendly way of production. Sustainability of the garments industry is also a burning issue, needed to adopt cleaner and improved technology and management for better environment.

1.23   Non-popularity of green fashion

The reasons for non-popularity of green fashion have to be looked into two different lines—firstly, why textile industry is not prone to green fashion and secondly, why the customers are not giving preference to green fashion. Those concerned with textile industry are quite aware that this industry is the most ecologically harmful industry in the world. The eco-problems in textile industry occur during some production processes and are carried forward right to the finished product. In the production process of bleaching and then dyeing, the subsequent fabric makes a toxin that swells into our eco-system. During the production process controlling pollution is as vital as making a product free from the toxic effect. The control over this industry to have safety measures for checking the pollution is not very effective.

Textile industry owners are also quite aware that the utilisation of rayon for clothing is adding to the fast depleting forests in India. So far the textile industry is concerned for non-popularity of green fashion; the major drawback in going green for it is the cost. In some cases, using green products and materials will cost much more than using conventional materials. They fear that the increase in cost will make clothes more costly and beyond the reach of ordinary middle class families in India compels them for using conventional material. Moreover, time is another area where going green is a disadvantage according to textile manufacturers.

The textile industry that goes green needs to spend time researching the best ways to make the transition to green. In addition, the industry needs to locate sources of green material and green products and make sure that personnel are properly instructed in the use of the new products. Though the fear in textile industry is quite realistic, still the initiatives are not many. Garment manufacturers in Punjab, who initiated an awareness campaign on the ill effects of cotton cultivated with toxic chemicals, have started launching exclusive garments made of organic cotton. Ludhiana based garment manufacturer namely, Venus Group marketing its products under UV&W' brand announced its foray into the domestic market with the launch of first ever organic cotton-based apparels in 2008. These organic garments are certified by the Global Organic Textile Standards and Organic Exchange standards. Although the organic cotton yarn prices are 40% higher than the regular cotton, the company claims to have reasonably priced its products. The initial response which was quite slow has gained momentum within few years touching 40 million dollars recorded turnover in one year. It is worth noting that Venus Group, which was earlier exporting its garments to leading brands such as Gap, Wal-Mart and Carrefour, is today offering organic clothing for customers in India.

Previously, this textile industry has done something which has not been good for environment. These are some corrective measures taken by this industry under increasing pressure of socially aware people who want to opt for skin-friendly garments. This shows that the fear about cost and its acceptability by the customers is not always true. Another reason for non-popularity of green fashion among the customers is not only the lack of awareness, but also the non-availability of eco-friendly clothes. We have already seen that more than three-fourth selected customers are not aware about eco-friendly clothes or green fashion, environmental hazards posed by textile industry and they also don't know the difference between eco-friendly and ordinary clothes which are not eco-friendly. None of the shops or company outlets displays green fashion separately and prominently for the convenience of the customers. Even the exclusive shops or outlets of green fashion are totally absent and are still to make their appearance even in metropolitan cities.

1.24   Ways to popularise green fashion

Worldwide evidence indicates people are concerned about the environment and are changing their behaviour accordingly. As a result there is a growing market for sustainable and socially responsible products and services. Though the green trend is more discernible in the developed countries, it has slowly started gaining ground in the developing countries as well. In India too, concern for the environment has considerably heightened in recent years and this is evident from the increasing enactment of environmental legislations and judicial activism. Business firms have also started turning green and embracing green marketing practices to conform to green pressures and environmental legislation. In fact, with the threat of global warming and ecological degradation looming on our heads, eco-friendly products including green fashion have to find a favour amongst the coming generations in India also. However, it is not necessary that increased awareness about environmental degradation may change consumers' behaviour for going green. For example, a study conducted by Banumathi Mannarswamy has shown that although the customers in a city like Coimbatore are aware of environmental problems and green products in the market, but their attitudes and the behaviour towards the green purchase has not improved. In fact, till sometime back, the term "eco-friendly clothing" or green fashion was completely alien to India. Though the concept has long been popularised in the West, India caught it not too long back and hence its sustainability in India requires systematic campaign in its favour without which neither the textile industry nor the customers in large number will go for green fashion. The results of this study show that customers can be motivated to go for green fashion by increasing their level of awareness about the merits of eco-friendly clothes on the one hand and environmental hazards of the production process involved in non-eco-friendly fibres. This can increase the demand for green fashion in the market which may put pressure on textile industry to opt more for eco-friendly fibres.

Clothing labels generally reveal what fibres are used to make a garment and how to clean it. However, those labels don't outline all the chemical finishes applied to the garment or the environmental impact of the manufacturing process. If it is made mandatory for the textile industry to specify clearly and prominently the adverse effect and hazards of environment on non-eco-friendly clothes, people will become aware themselves and don't go for clothes made of such fibres. There is need to develop environmentally-responsible or green marketing in India and intensively campaign for it. Green marketing is a business practice that takes into account consumer concerns about promoting preservation and conservation of the natural environment. Green marketing campaigns highlight the superior environmental protection characteristics of a company's products and services, whether those benefits take the form of reduced waste in packaging, increased energy efficiency in product use, or decreased release of toxic emissions and other pollutants in production. This campaign along with the ready availability of eco-friendly clothes and their affordable cost may help sustaining green fashion in India. This study has highlighted the fact that health conscious customers are even ready to pay more for green fashion in their own interest and in the larger interest of the nation. It should be insensibly advertised that the green fashion is environmentally friendly clothing. Audio-visual media of mass communication like television and print media should be used for advertising green fashion. It should be brought to the notice of people at large that fibres making up the green fashion are derived from the Earth's natural environment without depleting limited natural resources.

They inflict minimal harm to human and environmental health as they implement improved manufacturing measures and eliminate contaminated waste to a large extent. An overwhelming majority of the customers showed their awareness about environmental protection and felt a strong need for taking stringent measures to save the environment. As they are not aware about the adverse impact of chemicals and dyes used in textile industries on their lives and environment, they are opting for clothes made of non-eco-friendly fibres. The increased awareness among people can help sustaining green fashion. It must be remembered that the future belongs to clothing that is kind to both the environment and people, and labels with high standards in both respects. Farmers should be encouraged to grow organic cotton and other eco-friendly fibres. As the growth process of the harvested fibres does not involve chemicals, harvesting such fibres will reduce the cost.

Government should purchase these fibres for onward supply to textile industry and give incentive to textile industry for giving due cast of these fibres to the farmers. Checks should also be made on synthetic fabrics, such as polyesters, nylons, and acrylics which have adversely affect human body and help in degrading the environment. Animals are also another source of natural fibres. Fibres derived from various animals such as rabbit, sheep, llama, goats, etc., is also used for green fashion and such animals also need legal protection for availability of fibre in large quantity. Farmers should also be encouraged for helping in such fibres. In the light of empirical evidence, it is suggested that government should organise informative programs to make the public aware about need for green clothing, processes and dyes used in other clothes that adversely affect the nature, their processing with heavy chemical agents which are not only harmful and have enduring effects on environment but also on the health of people. In the larger interest of a nation and its people, government should provide liberal subsidies to units engaged in manufacturing green clothes and should adopt stringent measure to control the use of harmful chemicals and dyes in textile industry. In order to make the textiles totally environment friendly, not only the final product to be used by the consumer should be eco-friendly, but the production technology, packaging and disposal after use should also be eco-friendly. Therefore, the production ecology, user ecology and disposal ecology must be taken into consideration. Government should discharge its duty in convincing textile industry to consider all these factors and if the industry does not fall in line, stringent legal measures should be adopted to have control on the industry in the larger interest of the people and the environment.

1.25   References

1. Allwood, J.M., Laursen, S.E., Malvido de Rodriguez, C.M., and Bocken, N.M.P. (2006), *Well dressed? The present and future sustainability of clothing and textiles in the United Kingdom*, Cambridge: Institute for Manufacturing, University of Cambridge. Available from: www.ifm.eng. cam.ac.uk.
2. Al-Tuwaijri, S.A., Christensen, T.E., and Hughes II, K.E. (2004), The relations among environmental disclosure, environmental performance, and economic performance: a simultaneous equations approach, *Accounting, Organizations and Society*, 29(5–6): 447–471.
3. Aspiras, F.F. and Manalo, J.R.I. (1995), Utilization of textile waste cuttings as building material, *Journal of Materials Processing Technology*, 48(1): 379–384.
4. Baker, S. (2000), The EU: Integration, Competition and Growth and Sustainability, W. Lafferty, and J. Meadowcroft, Eds. *Implementing Sustainable Development*, Oxford: Oxford University Press.
5. Bechtold, T., Turcanu, A., Ganglberger, E., and Geissler, S. (2003), Natural dyes in modern textile dye houses—how to combine experiences of two centuries to meet the demands of the future?, *Journal of Cleaner Production*, 11(5): 499–509.
6. Bogoeva-Gaceva, G., Avella, M., Malinconico, M., Buzarovska, A., Grozdanov, A., Gentile, G., and Errico, M.E. (2007), Natural fiber eco-composites, *Polymercomposites*, 28(1): 98–107.
7. Borchardt, M., Wendt, M.H., Pereira, G.M., and Sellitto, M.A. (2011), Redesign of a component based on ecodesign practices: environmental impact and cost reduction achievements, *Journal of Cleaner Production*, 19(1): 49–57.
8. Boujarwah, F.A., Mogus, A., Stoll, J., and Garg, K.T. (2009), Dress for success: automating the recycling of school uniforms. In: *CHI'09 Extended Abstracts on Human Factors in Computing Systems*, pp. 2805–2810.
9. Briga-Sá, A., Nascimento, D., Teixeira, N., Pinto, J., Caldeira, F., Varum, H., and Paiva, A. (2013), Textile waste as an alternative thermal insulation building material solution, *Construction and Building Materials*, 38: 155–160.
10. Brinberg, D. and McGrath, J.E. (1985), *Validity and the Research Process*. Beverly Hills: Sage Publications.
11. Bryman, A. (2004), *Social Research Methods (2nd ed.)*. Oxford: Oxford University Press.
12. Caniato, F., Caridi, M., Crippa, L., and Moretto, A. (2012), Environmental sustainability in fashion supply chains: an exploratory case based research. *International, Journal of Production Economics*, 135: 659–670.

13. Çetinkaya, E., Rosen, M.A., and Dinçer, İ. (2012), Life cycle assessment of a fluidized bed system for steam production, *Energy Conversion and Management*, 63: 225–232.

14. Chen, H.L. and Burns, L.D. (2006), Environmental analysis of textile products, *Clothing and Textiles Research Journal*, 24(3): 248–261.

15. Chi, T. (2011), Building a sustainable supply chain: An analysis of corporate social responsibility (CSR) practices in the Chinese textile and apparel industry, *Journal of the Textile Institute*, Vol. 102:, pp. 837–848.

16. Cormier, D. and Magnan, M. (2003), Environmental reporting management: a continental European perspective, *Journal of Accounting and Public Policy*, 22(1): 43–62.

17. COTANCE (2012), *Social and Environmental Report - the European leather industry*, Available from: http://cotance.com/socialreporting/european-reporting/european-ser. html.

18. Council for Textile Recycling. (1998), *Don't overlook textiles*, Available from: http:// textilerecycle.org.

19. Crowther, M.A. and Cook, D.J. (2007), Trials and tribulations of systematic reviews and meta-analyses, *Hematology*, pp. 493–497.

20. Curwen, L.G., Park, J., and Sarkar, A.K. (2013), Challenges and solutions of sustainable apparel product development a case study of Eileen Fisher, *Clothing and Textiles Research Journal*, 31(1): 32–47.

21. Mahajan, S. (2011), Origin and functions of clothing: A note, *Journal of National Development*, 24(1): 239–244, ISSN 0972-8309.

22. Elkington, J., Julia H., and Joel M. (2013), *The Green Consumer*, Penguin Group USA, ISBN: 0140177116.

23. Jain, S.K. and Kaur, G. (2004), Green marketing: An Indian perspective, *Global Business Review*, 5(2): 187–205, ISSN 0972-1509.

24. Mannarswamy, B. (2011), A study on the environmental awareness and the changing attitude of the students and public in Coimbatore towards green products, *Research Journal of Social Science and Management*, 1(7): 75–84, ISSN 2251-1571.

25. Noor, A.A.S., Norhayati, Z., Ahmad, J., and Mohammed , S., and Mohammed, A. (2012), Green supply chain management: a review and research direction, *International Journal of Managing Value and Supply Chains (IJMVSC)*, 3.

26. Nimawat, D. and Namdev, V. (2012), An overview of green supply chain management in India, *Research Journal of Recent Sciences*, 1, 77–82

27. Parray, S.H. and Kadri, S.M. (2007), Supply chain management in healthcare sector - Role of logistics, *Indian Journal for the Practising Doctor*, 4, 42–51

28. Denyer, D. and Neely, A. (2004), Introduction to special issue: innovation and productivity performance in the UK, *International Journal of Management Reviews*, 5/6: 131–135.

29. Dickson, M., Loker, S., and Eckman, M. (2009), *Social Responsibility in the Global Apparel Industry*, New York: Fairchild.

30. Domina, T. and Koch, K. (1997), The textile waste lifecycle, *Clothing and Textiles Research Journal*, 15(2): 96–102.

31. Dubey, R. and Bag, S. (2013), Exploring the dimensions of sustainable practices: An empirical study on Indian manufacturing firms, *International Journal of Operations and Quantitative Management*, 19(2): 123–146.

32. Easton, J. (2007), Supply chain partnerships for sustainable textile production, M. Miraftab ed., *Ecotextiles: The way forward for sustainable development,* pp. 50–57, Cambridge, England: Woodhead Publishing.

33. Eisenhardt, K.M. (1989), Building theories from case study research, *Academy of Management Review,* 14(4): 532–550.

34. Goldbach, M., Seuring, S., and Back, S. (2003), Co-ordinating sustainable cotton chains for the mass market, *Greener Management International,* 43: 65–78.

35. Goworek, H. (2011), Social and environmental sustainability in the clothing industry: a case study of a fair trade retailer, *Social Responsibility Journal,* 7(1): 74–86.

36. Gwilt, A. and Rissanen, T. (2011), *Shaping Sustainable Fashion: Changing the Way We Make and Use Clothes,* London: Earthscan. 65–78.

37. Hackston, D. and Milne, M.J. (1996), Some determinants of social and environmental disclosures in New Zealand companies, *Accounting, Auditing & Accountability Journal,* 9(1): 77–108.

38. Hayes, L.L. (2001), Synthetic textile innovations: Polyester fiber-to-fiber recycling for the advancement of sustainability, *AATCC Review: The Magazine of the Textile Dyeing, Printing, and Finishing Industry,* 11(4): 37–41.

39. Holsti, O. (1969), *Content analysis for the social sciences and humanities,* Reading: Addison-Wesley. 114–128

40. Jacques, J.J. and Guimarães, L.B. (2012), A study of material composition disclosure practices in green footwear products, *Work: A Journal of Prevention, Assessment and Rehabilitation,* 41: 2101–2108.

41. Jenkins, H. and Yakovleva, N. (2006), Corporate social responsibility in the mining industry: Exploring trends in social and environmental disclosure, *Journal of Cleaner Production,* 14(3): 271–284.

42. Jose, A. and Lee, S.M. (2007), Environmental reporting of global corporations: a content analysis based on website disclosures, *Journal of Business Ethics,* 72(4): 307–321.

43. Kalliala, E. and Talvenmaa, P. (2000), Environmental profile of textile wet processing in Finland, Environmental profile of textile wet processing in Finland, *Journal of Cleaner Production,* 8(2): 143–154.

44. Koch, K. and Domina, T. (1999), Consumer textile recycling as a means of solid waste reduction, *Family and Consumer Sciences Research Journal,* 28(1): 3–17.

45. Kumar, M., Aravindhan, R., Sreeram, K., Rao, J., and Nair, B. (2011), Green chemistry approach in leather processing: A case of chrome tanning, *Journal of the American Leather Chemists Association,* 106(4): 113–120.

46. Levy, Y. and Ellis, T.J. (2006), A systems approach to conduct an effective literature review in support of information systems research, *International Journal of an Emerging Trans Discipline,* 9: 181–212.

47. MacCarthy, B.L. and Jayarathne, P.G.S.A. (2012), Sustainable collaborative supply networks in the international clothing industry: a comparative analysis of two retailers, *Production Planning & Control,* 23(4): 252–268.

48. Maignan, I. and Ralston, D.A. (2002), Corporate social responsibility in Europe and the US: Insights from businesses self-presentations, *Journal of International Business Studies,* 33(3): 497–514.

49. Miles, M.B. and Huberman, A.M. (1984), *Qualitative Data Analysis,* Beverly Hills, CA: Sage Publications. 1–6

2

Implementation of sustainability initiatives for green production in apparels

S. Senthilkumar[1] and P. Vinayagamurthi[2]

[1]Department of Handloom and Textile Technology, Indian Institute of Handloom Technology, Salem – 636 001 Email: thil.tex@gmail.com
[2]Department of Costume Design and Fashion, Sri Jayendra Saraswathy Maha Vidyalaya College of Arts and Science, Coimbatore – 641 005
Email:vmcostume@gmail.com

Abstract: Sustainable textiles are grown and created in an environmentally friendly way, using minimal chemicals. Due to chemicals are not used in sustainable textiles, there are less health problems that are associated with chemicals such as headaches, allergies, skin irritation, and respiratory problems. For a textile to be sustainable, it has to be made from a renewable resource, it has to have a good ecological footprint and it should not use any chemicals in the growing and processing of it. Sustainability is defined as the design of human and industrial systems to ensure that mankind's use of natural resources and cycles do not lead to diminished quality of life—either to losses in future economic opportunities or to adverse impacts on social conditions, human health, and the environment.

Keywords: sustainable textiles, ecological footprint, renewable resource, environment

2.1   Introduction

Sustainability is the major concern in the age of modern world. For textile and apparel sector, this has been a burning issue for many related concerned organisations. Over the past few years, increasing awareness of the environmental and social concerns surrounding the fashion industry have led to a rise in the implementation of sustainability initiatives. Sustainability issues in the textile and apparel industry have received great attention. With geographically long and complex global production networks, as well as the dual pressure for cost and lead time, implementing sustainability in textile and apparel supply chains is challenging The process of turning raw materials into finished garments has significant negative environmental and social implications, including air and water pollution and exploitation of human resources, especially where production is outsourced to lower labour cost countries. The triple bottom line approach suggests that companies should consider social and environmental performance, not only financial performance, in their business operations. Furthermore, with increasing awareness of sustainability, there is evidence that some consumers are willing to pay more for sustainable textile and apparel products.

2.2   Sustainability definition

The most suitable definition of sustainability recommended by the world Commission on Environment and Development is "meet the needs of the present without compromising the ability of future generation to meet their needs and desires" (World Commission on Environment and Development).

2.2.1   Sustainability features

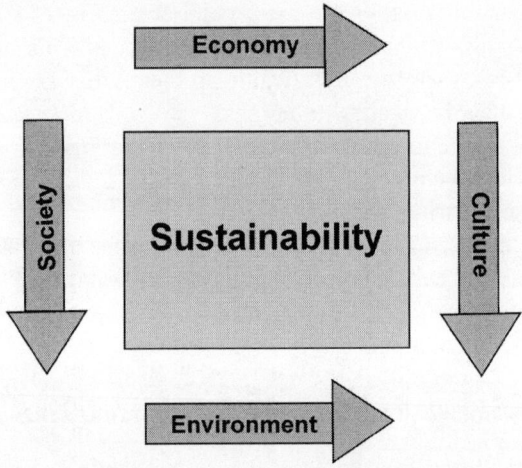

Figure 2.1. Sustainability features

Sustainability is working on four main features like economy, society, culture and environment (Figure 2.1). And the question for future is to make and fulfil the demands and desires by keeping in view about all the wastes and extra materials we are disposing off without thinking about upcoming new generation, future and environment; because we have to save the planet in future. To overcome and solve these issues, we need complete and efficient management systems along with resources to implement these systems.

In recent times sustainability is a leading characteristic of textile fashion products. Textile fashion companies are focusing more on sustainable products these days, so that they can meet the environmental and social aspects. For getting competitive advantage in fashion business the companies have to take care of social, political and economic issues, and they must be aware of current trends of the market. Sustainable fibres provide solution for the companies facing issues regarding environmental problems; these fibres are also favourable to meet the market demands of quality products these days. Environmental science defines sustainability as the quality of not being harmful to the environment or depleting natural resources and thereby supporting long-term ecological balance. The sustainable textile products can be defined as the textile products produced using raw materials, energy, various resources, and other ingredients which are derived from renewable resources that cannot be exhausted and consequently do not affect the next generation.

2.2.2   Role of sustainability in textile

Focusing on textile sustainable materials the term sustainability plays a vital role in textile processing and in all steps. As today requirement is to produce goods that are eco-friendly and have ability to serve the customer as well as environment and also support the economy of the company. Textile manufacturing industries have impact on environment in different ways i.e., use of water in growing cotton, energy consumption in all processes and use of chemicals and materials in the processing. Apart from this, the industry is always trying to find some solutions for problems like pollution, health, global warming and environmental issues. To answer and overcome these issues, technologies and innovative strategies are there. So the need is to create awareness among the company management and people. And it does not affect the company cost to implement these steps and the money will be back from the efficient supply chain and competitive advantage in the market.

Efficiency is fundamentally linked with sustainability. Making "more with less" is the heading step towards sustainability. If different companies and brands produce the goods with quality and using less energy and less inputs, than these companies will grow and make profit naturally, and it will also be beneficial for environment.

2.3   Sustainable textile fibres

Natural fibres are mostly considered as more sustainable and synthetic fibres are considered less sustainable. This assumption is based on the fact that natural fibres production needs less consumption of resources than synthetic fibres and also synthetic fibres have impact on people and environment. But the fact involves much more, though production of synthetic fibres needs lot of resources for their production, but the cotton cultivation has also high impacts, for growing cotton large amounts of pesticides, fertilisers and water is needed.

For producing 1 kg of cotton needs 8000 litres of water, while producing 1 kg of polyester uses less water, however, it needs twice energy for its production when we produce the same amount of cotton. Organic cotton or low chemical cotton is a sustainable alternative of conventional cotton as it has social and environmental benefits, because organic cotton production is characterised with no use of synthetic pesticides, fertilisers and less water consumption. The fibres also have other social and ethical impacts, the emission of carbon while producing synthetic fibres is a major issue these days, so there is need of carbon neutral fibres e.g., plant fibres like bamboo and Lyocell—these fibres absorb the same amount of carbon dioxide gas from environment during their growth as they release during their production cycles, thus helping in keeping the atmosphere clean. Synthetic fibres consume lot of oil during their production, so there is a shift from oil-based non-biodegradable synthetic fibres like polyester and nylon towards renewable and biodegradable synthetic fibres produced from natural resources like Lyocell and PLA (poly lactic acid). So these natural plant fibres and synthetic fibres made from natural resources are naturally eco-friendly, less resource consuming, recyclable and sustainable.

2.4 Classification of fibres

2.4.1 Organic cotton

Conventional cotton is very environmentally unfriendly as the extensive use of pesticides and insecticides used when growing the cotton cause pollution and also ill health. Organic cotton however is grown without the use of chemicals, making it much more environmentally friendly (Figure 2.2).

2.4.2 Hemp

Pesticides or insecticides are not needed when growing hemp and hemp actually improves the condition of the soil that it is grown in. It is also drought resistant and can be grown in most climates. The fabric can be made from the hemp plant without using toxic chemicals and it can be processed locally, reducing the costs and pollution associated with transport.

2.4.3 Bamboo

As a plant, bamboo is very fast growing, helps to improve the quality of the soil, and can help to rebuild eroded soil. It is very sustainable. Bamboo fabrics can be made mechanically or chemically. Because strong solvents are used in the chemical method, it is not considered a sustainable way to create fabric. However, there are newer manufacturing methods that are environmentally friendly. Look for a label from an organic or sustainable certification body.

2.4.4   Soya

Soya cloth is made from a by-product that occurs during the food manufacturing of the soya bean. The fabric is soft, drapes well, and is comfortable. Look for soya cloth that is certified organic.

2.4.5   Wool

Wool can be an environmentally friendly fabric with some conditions—the animals need to be treated well and live in humane conditions. The sheep manure should not enter the water supply. Another consideration is how the wool is manufactured—environmentally friendly wool will not use bleach or chemical dyes.

2.4.6   Pina fabric

Pineapple leaves are used to obtain Pina, a textile fibre that is used to make fabrics. The pine fibres are extracted from the pineapple leaves by hand scraping, decortications or retting.

2.4.7   Stinging nettle fibre

This fibre is obtained from the Brennessel plant which is naturally resistant to vermin and parasites. It can be grown without pesticides and herbicides and with very little fertilisation as the minerals do not get leached out of the ground. They can be mixed with organic cotton and spun into yarn. Nettle fibre is stronger than cotton and finer than linen fibre. They can be made into a wide range of woven as well as knitted fabrics. Due to its fine weft and glossy look, nettle fabric was very popular in middle ages but lost its position to inexpensive cotton. Now again, it is becoming popular as sustainable alternative to cotton.

2.4.8   Milk protein fibre

These fibres are used to make yet another and eco-friendly yarn—milk yarn. Milk is dewatered, i.e., all the water content is taken out from it and then skimmed. With the help of bio-engineering technique, a protein spinning fluid is made. Wet spinning process converts this fluid into high-grade textile fibre. The skin friendly milk yarn goes to make glossy and luxurious fabrics similar in appearance to silk fabrics that have antibacterial and antifungal properties too. Their hygroscopic character makes them one of the finest moisture management fabrics. They can be blended with a number of fibres to get many characteristics—blend them with bamboo to get cool fibre and with wool fibre to have a thermal protective fibre.

2.4.9   Banana fibre

The banana fibre is extracted by hand stripping and decortications. Thus, it is 100% eco-friendly fibre. This fibre looks like bamboo fibre and ramie fibre. It is strong, shiny, lightweight and biodegradable. It can even absorb moisture very efficiently. Banana fibres were used for making ropes and mats till recent past. With its many qualities getting popular, the fashion industry is also fast adopting this fibre for making various fashion clothing and home furnishings.

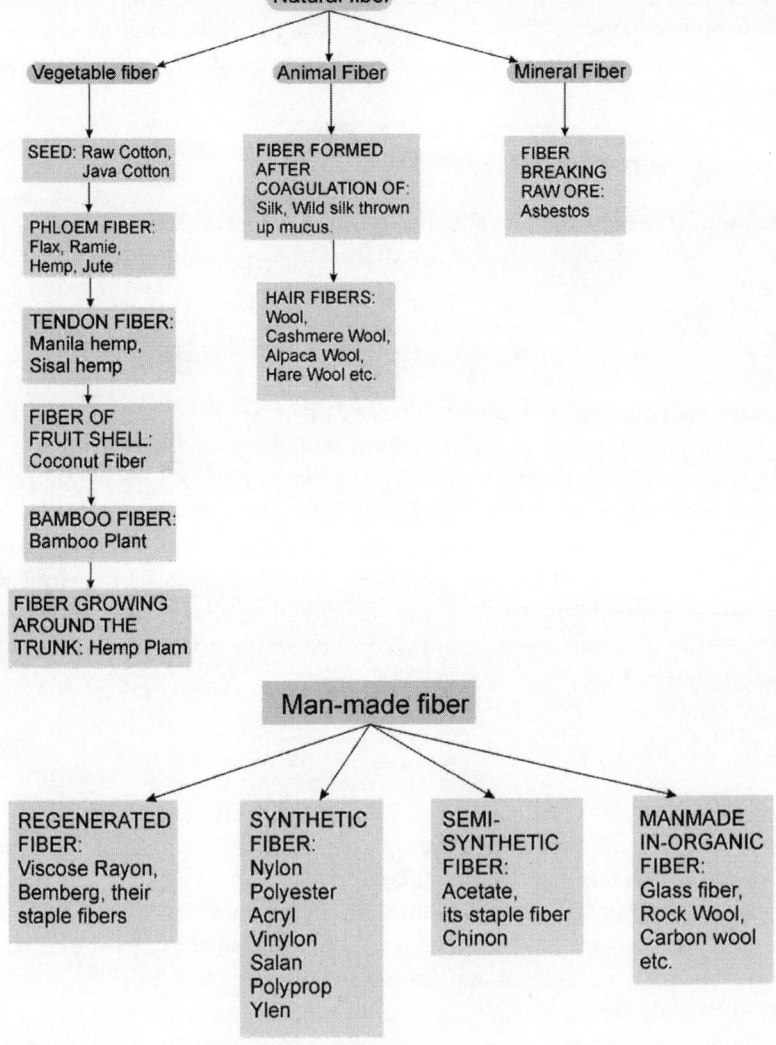

Figure 2.2. Classification of fibres

2.5   Key factors in organic cotton development

There are several factors (Figure 2.3) that played a key role in rapid and incessant development of organic cotton products in the global market, some of the main factors are described below.

- Changing life style of consumers
- Companies making their business strategies more sustainable
- Access to knowledge about organic program development.

2.5.1   A systematic approach for growth of organic cotton

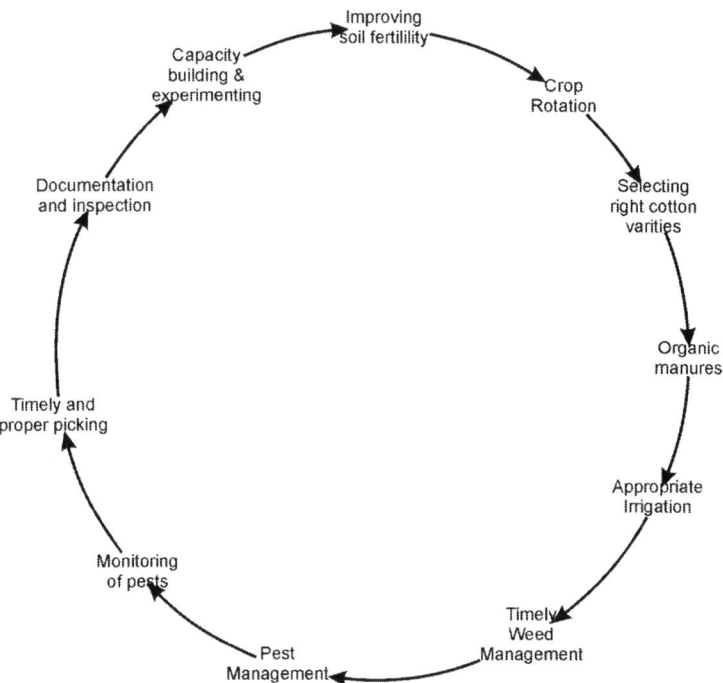

Figure 2.3. Systematic approach for growth of organic cotton

2.5.2   Advantages of organic cotton

The advantages of organic cotton are tabulated in Table 2.1.

Table 2.1. Advantages of organic cotton.

S. No.	Resources	Conventional cotton	Organic cotton
1	Environment	• Water pollution • Loss of biodiversity • Adverse changes in water balance • Pollution of soil and air • Pesticides killing beneficial insects	• Improved water utilisation • Increased biodiversity • Soil and air are hygienic • Eco-balance between pests and insects
2	Social	• Health problems in regions where regulatory systems are week • Poisoning and causalities due to extensive use of pesticides	• Use of local varieties and resources • Helpful for low income families due to more premium
3	Economy	• Resource consuming • High production cost • No alternative crops	• Less resource consumption • Lower production costs • Niche product • More revenue for farmers
4	Food	• Pesticides entering human food through cottonseed oil • Contamination of meat and milk from animals fed on cotton products	• No danger of contamination of edible items originated from cotton source
5	Agricultural	• Reduced soil fertility • Poor irrigation, contamination fields becoming barren	• Increased soil fertility • Crop rotation maintains soil structure
6	Health	• Chemicals remained in final product cause health problems • Chronic diseases (cancer, infertility, weakness, illness)	• No use of pesticides, or chemicals that saves the farmer and surroundings from chronic diseases

2.5.3   Disadvantages of organic cotton

Besides of the enormous benefits of organic cotton production here are some problems associated with organic cotton that act as a barrier in development of organic cotton.

2.5.3.1   Productivity

The productivity of organic cotton is less than the conventional cotton, in conventional cotton production the farmers use growth fertilisers for higher productivity and are reluctant to grow organic cotton.

2.5.3.2   Cost

The growth of organic cotton in a field of conventional cotton needs a transition or conversion period, during this period farmers observe the organic cotton standards but they are not able to sell this cotton as organic cotton, this works as an obstacle to convert conventional cotton growth into organic cotton without financial assistance to farmer.

2.5.3.3   Cultivation

In conventional farming systems seeds are directly sown in the soil, while in organic cotton growing first mostly weeds are removed prior to sowing the seeds. So farmers feel easy to grow the conventional cotton.

2.5.3.4   Time

Growing organic cotton is a systematic approach that needs a lot time and there is a transition time of approximately 3 years to get organic cotton from conventional field so the farmers have not the organic premium in this time. Also organic cotton timely intervention during its growth, a farmer can naturally produce more crops with industrial methods.

2.5.4   The better cotton initiative (BCI) and best management practices (BMPs)

The BCI's BMP program is designed to improve the production practices of as many farmers as possible, rather than excluding farmers who do not employ practices consistent with the BCI's production principles. This approach enables more conventional farmers to transition into biological systems, thereby expanding the total area of land with BMP (BCI, 2008). The defining feature of the BCI's BMP is its emphasis on water management and labour in addition to reduced chemical use. BCI production principles include:

- Minimising the use and impact of pesticides;
- Optimised use of water and care for water availability;
- Conservation of natural habitats;
- Water must be extracted legally with neither groundwater nor water bodies used for irrigation being adversely affected;
- Soil management practices maintain and enhance the structure and fertility of the soil and include minimum tillage, cover crops and rotation crops;
- Nutrients are applied to minimise risk to the health of workers; labour rights include voluntary overtime with pay, no child labour, etc.
- Preservation of fibre quality is a priority.

2.5.5   Market bottleneck caused by organic cotton

Organic cotton is seen as the pinnacle of sustainability in the market and creates a bottleneck, restricting market incentives for farmers transitioning to biological systems (Figure 2.4).

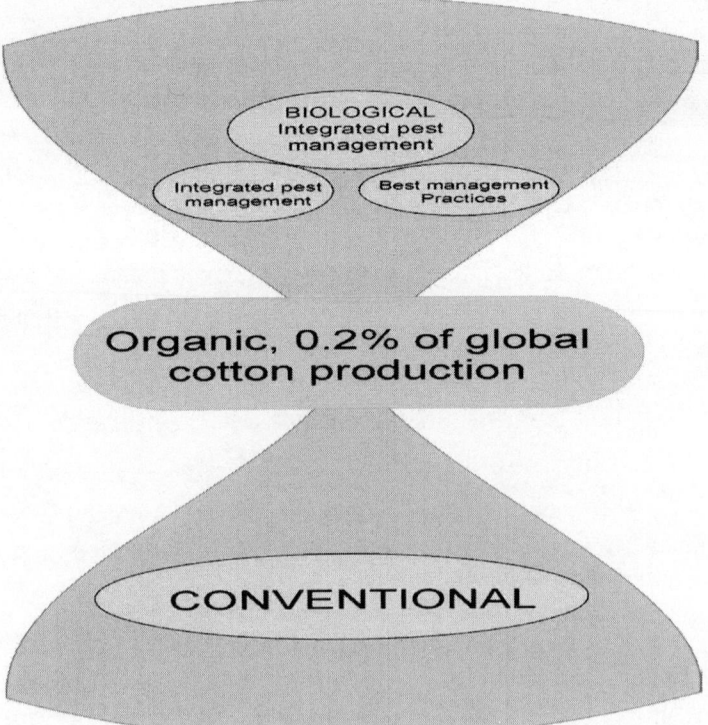

Figure 2.4. Market bottleneck caused by organic cotton

2.6   Expanded options for sustainable cotton

Different farm systems (‚numbers 1 to 6 shown in Figure 2.5) are often perceived as being in competition with each other, when they are in fact working towards the same goals. Organic production is one tool that is a stepping-stone to more sustainable practices in cotton, especially in areas of low chemical use. Additional biological systems broaden ecological goals through scalability and may be "sustainable" in and of themselves. GM (genetically modified) and BT (biologically transformed) cotton in particular may initially provide a stepping-stone to biological systems in highly degraded areas where chemical use and toxicity is high, but the risks of evolving genetic resistance in insects are real and acknowledged by everyone.

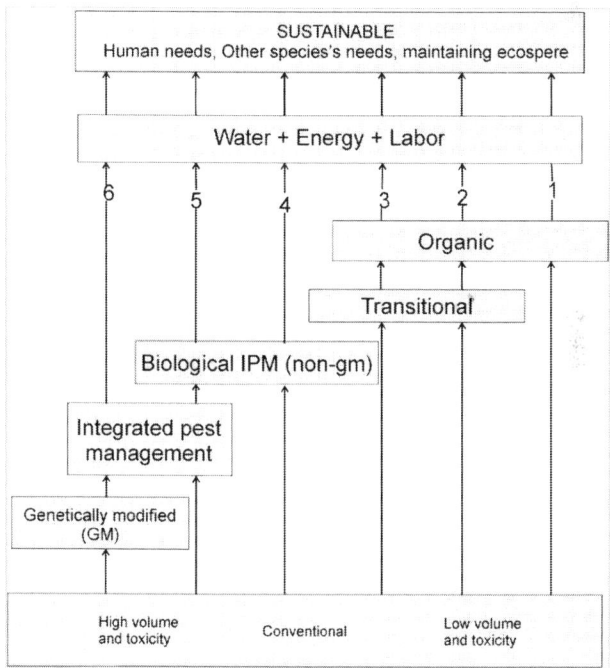

Figure 2.5. Expanded options for sustainable cotton

2.7   Methods of bamboo manufacturing

There are two methods of manufacturing of bamboo. They are as follows:

1. Mechanical method.
2. Chemical method.

2.7.1   Mechanical method

In this method "woody" part of bamboo plant crushed and treated with natural enzymes to break down the bamboo in to the soggy substance and then this material is combed-out into spun yarn. It is the most reasonable method of manufacturing bamboo yarn and the bamboo produced by this method is very rough like linen. The problem with this method is that it's a little expensive method and is not preferable by Chinese manufacturers. In reality all the bamboo products, bed sheets, garments, fibres, fabrics are made by China and only few companies of USA are using this type fibre or yarn in their products after importing the material from China (Figure 2.6).

Figure 2.6. Mechanical method bamboo manufacturing

2.7.2   Chemical method

In this method sodium hydroxide is used to crush the woody bamboo and then the crushed bamboo is dipped into caustic soda and it converted into cellulose material. In US market products like bamboo beddings, shirts and duvet covers and towels are made by this type of bamboo.

There is some misconception in people about sodium hydroxide, but it does not cause any harm to environment and health of workers. It is approve by Global Organic textile Standards (GOTS) and the soil association that it can be used in organic cotton processing and it cannot stay on bamboo fabric as residue and easily remove after washing and can be neutralise to non-toxic sodium salt. In spite of that sodium hydroxide have many uses, it is used in paper making, food preparation, soap manufacturing for washing and soft drink productions, chocolate and cocoa processing and thickening of ice cream.

2.7.3   Properties

Bamboo rich with mineral elements and have approx. 17 sorts of amino acid and 2, 4% of protein. Bamboo elements serves human being medically it shoots consist of germaclinium that is used to activate human. Bamboo shoots contain amino acids like Lys, Arg and Glu—Lys amino acid that help in growth and development of child. Bamboo shoots are richer in mineral elements and amino acid than rice, corn, vegetables and flour. Bamboo also contains 0.05% fat, 2.5% sugar content and the fibre length is 0.65–1.2% that was fit for human consumption. Bamboo contains different numerous elements like chromium, zinc, magnesium, manganese, nitrate, copper, etc.

- The original bamboo fibre is similar to ramie or bast fibres.
- Bamboo fibre is bit thinner and finer than ramie fibre.
- It has specialised and functional properties like anti-bacterial, anti-UV, deodorant and germicidal.
- According to research it has been approved that the properties of thinness and whiteness of bamboo fibre is better than the bleached viscose.
- Strong durability, tenacity and stability.
- Breathable and cool.

Japan textile inspections association tested the bamboo fibre fabric and washed those fabrics up to 50 washes and found that the fabric still possess the anti-bacterial properties approx. 70%. On the other hand, when apparels treated with chemical and anti-microbial materials they cause irritation and skin allergy. So the bamboo fibre used in all surgical items, bandages and mostly hospital products e.g., bed sheets, gowns, masks, nurses wear, socks, cloths, underwear, decorative items, etc.

2.8   Wool

It is largely the fibre diameter of the wool that determines the value and the end uses of the harvested fleece. British breeds produce mostly coarser quality wool (fibres are around 30 micrometres (or microns) in diameter or more). This wool is highly suited for products such as carpets, blankets and hand-knitting yarns. Finer wools (with fibres of 20 microns in diameter or less) produce soft and luxurious fabrics that are highly suited to next-to-skin wear, and that can be used for applications that range from active sportswear to elegant evening wear. For fine wool garments it is important that wool is uniformly white and contains no dark fibres that would limit the range of garment colours that can be produced. This is less important for the coarser fibres where naturally coloured fibres can be tolerated or may even enhance the "natural" character of the products.

Global consumption of textiles is large. The textile industry has grown, not only in line with global human growth, but also as a result of per capita consumption. Around 1900, humans used around 2 kg of textiles per head, whereas by 2010 consumption will exceed 10 kg per head. As with most other consumable goods, per capita consumption is weighted to the wealthiest countries, with the USA and UK consuming as much as ten times the quantity of textiles per head as in developing countries. The environmental problems are associated not only with resources used in the initial manufacture of the textiles but also in the use of resources to clean and dispose of the garments. Most textile garments finish their life in landfill, a diminishing resource in most developed countries. It has been estimated that 1.2 million tonnes/year of textiles are sent to landfill in the UK, while 2 million tonnes/year are sent in Japan. This mass is split evenly between clothing and other textile products. There are increasing calls for greater recycling of textiles as this consumes much less energy than the manufacture of new textiles; however, problems with identifying useful end products for all of this recycled material remain difficult. Even in the UK which has reasonably well developed recycling systems, only around 25% of garments are recycled.

A second estimate for the combined waste from clothing and textiles in the UK is about 2.35 million tonnes (38 kg/person). Of the 330,000 tonnes of recovered textiles, 200,000 tonnes are exported while 100,000 tonnes are recycled within the UK. These recovered clothes are given to the homeless, sold in charity shops or sold in developing countries in Africa, the Indian sub-continent and parts of Eastern Europe. Over 70% of the world's population use second-hand clothes. Incineration is used for 10% of clothing while 60% is sent to landfill. The EU has indicated that there is likely to be legislation to encourage greater recycling of textiles, initially in the UK and France, but then across the EU. Because wool is a relatively expensive fibre, there is higher demand for discarded wool garments and processing wastes which are sold to specialist firms for fibre reclamation to make yarn or fabric. Incoming material is sorted into type and colour to minimise re-dyeing. The material is shredded into "shoddy" (fibres). Depending on the end uses of the yarn, other fibres are chosen to be blended with the shoddy by carding before spinning. Products may be garments, felt and blankets. In anaerobic landfills, wool, cotton and other natural fibres degrade rapidly while oil-based textiles degrade extremely slowly, but all potentially produce methane. Methane is a potent greenhouse gas and unless it is recovered, the global warming potential is greater than from incineration.

2.8.1   Life cycle assessment (LCA)

In 2006, the Commonwealth Scientific and Research Organization (CSIRO) and the Australian wool industry began a small project to perform a preliminary LCA on Australian wool. Other wool-producing countries are performing partial studies. Australian wool is used for the production of many types of garments and processing occurs in many different countries. In order to simplify the study and to allow identification of the major environmental pressure points, three typical Australian wool supply chains manufacturing common apparel products were selected.

- Fine wool grown in a high-rainfall climate by a specialist producer, processed in Italy into lightweight next-to-skin knitted garments;
- Medium micron wool produced on mixed enterprise farms, processed in Asia into men's suits;
- Slightly coarser wool, produced in Australia's arid pastoral zone, processed in Asia into an outerwear knit.

2.9   Different fibres impacts on the environment

European NGO with a mission to make the sustainable fashion a common practice, published a study in which the environmental impact of the production of several fibres is benchmarked. The Table 2.2 below summarises the results of the study. The fibres under Class A are believed to be the most environmentally friendly. This classification is not only based on water use, but also on energy use, land use, the use of non-renewable resources and the use of hazardous chemicals.

Table 2.2. Different fibres impacts on the environment.

S. No.	Class A	Class B	Class C	Class D	Class E	Unclassified
1	Recycled cotton	TENCEL® (Lenzing Lyocell product)	Conventional hemp	Virgin polyester	Conventional cotton	Silk
2	Mechanically recycled nylon	Organic cotton	Ramie	Poly-acrylic	Virgin nylon	Organic wool
3	Mechanically recycled polyester	Chemically recycled polyester	PLA	Modal® (Lenzing viscose product)	Spandex (Elasthane)	Leather

4	Recycled wool	In conversion cotton			Bamboo viscose	Natural bamboo
5	Organic hemp	Chemically recycled nylon	Conventional flax (Linen)		Wool	Acetate
6	Organic flax (Linen)	CRAiLAR® Flax			Generic viscose	Cashmere wool
7		Monocel® (Bamboo Lyocell product)			Cupram-monium rayon	Alpaca wool
8					Rayon	Mohair wool
9	More sustainable			Less sustainable		

2.9.1   Man-made fibres

Man-made fibres account for 68% of fibres used worldwide, and 75% of those processed in Europe. The World production was around 53 million tonnes in 2010 and the European production was 3.8 million tonnes. Their principal end-use is in clothing, carpets, household textiles and a wide range of technical products—tyres, conveyor belts, fillings for sleeping bags and cold-weather clothing, filters for improving the quality of air and water in the environment, fire-resistant materials, reinforcement in composites used for advanced aircraft production, and many more. Fibres are precisely engineered to give the right combination of qualities required for the end-use in question—appearance, handle, strength, durability, stretch, stability, warmth, protection, easy care, breathability, moisture absorption and value for money. In many cases, they are used in blends with natural fibres such as cotton and wool. Man-made fibres come in two main forms.

- Continuous filament, used for weaving, knitting or carpet production;
- Staple, discontinuous lengths of fibre which can be spun into yarn uses such as fillings or non-wovens.

2.9.1.1   Man-made fibre classification

Man-made fibres are classified into three classes. They are as follows:

- Those made from natural polymers,
- Those made from synthetic polymers and
- Those made from inorganic materials.

2.9.1.2   Fibres from natural polymers

The most common natural polymer fibre is viscose, which is made from the polymer cellulose obtained mostly from farmed trees. Other cellulose-based fibres are Lyocell, Modal, Acetate and Triacetate. Less common natural polymer fibres are made from rubber, alginic acid and regenerated protein.

2.9.1.3   Fibres from synthetic polymers

There are many synthetic fibres, for example, organic fibres based on petrochemicals. The most common are polyester, polyamide (often called nylon), acrylic and modacrylic, polypropylene, the segmented polyurethanes which are elastic fibres known as elastanes (or spandex in the USA), and speciality high-tenacity fibres such as the high performance aramids and UHMwPE (Ultra High Molecular weight Polyethylene).

2.9.1.4   Fibres from inorganic materials

The inorganic man-made fibres are made from materials such as glass, metal, carbon or ceramic. These fibres are very often used to reinforce plastics to form composites.

2.9.2   Renewable raw material

Most of the man-made fibres (MMF) are made out of synthetic polymers where the feedstock is oil-based. However, some MMF are also based on renewable resources. Viscose, Lyocell and Modal are typical and important MMF which are based on wood as the renewable resource. Cellulose is the world's most important biopolymer by far. About half of the global biomass consists of cellulose, being an amazing unique biopolymer. Wood is converted into pulp, and pulp into fibres. The pulp and fibre industry is part of the natural carbon cycle.

Eucalyptus Tencel—Tencel Lyocell is produced exclusively from the wood pulp of Eucalyptus trees certified by the Forestry Stewardship Council (FSC), and the fibre carries the Pan-European Forest Council (PEFC) quality seal. Eucalyptus is woody and therefore needs energy input to convert it into a soft fibre suitable for clothing. The Eucalyptus is reduced down then reformed into a spinable fibre. This is done in a process with similar principles as other semi-synthetic natural fibres, such as viscous bamboo fabric. The process used to make Eucalyptus Tencel is much more eco-friendly; it is simply the most environmentally friendly man-made cellulosic fibre available today.

2.9.3   Biodegradable and sustainable fibres

More and more textile researchers, producers and manufacturers are looking to biodegradable and sustainable fibres as an effective way of reducing the impact textiles have on the environment. The emphasis in biodegradable and sustainable fibres is on textiles that are beneficial by their biodegradation and come from sustainable sources. Natural polymers are both biodegradable and sustainable. But research is still on-going to develop new synthetic polymers/ fibres derived from renewable sources. For a material to fulfil the "cradle to grave" sustainability requirement, it must be both derived from a renewable source and be degradable (Figure 2.7).

- a. The biodegradable polymers can be classified into three main categories.
- b. Natural polysaccharides and biopolymers; e.g., cellulose, alginate, wool, silk, chitin and soya bean protein.
- c. Synthetic polymers, particularly aliphatic polyesters e.g., poly (lactic acid), Poly(ε-caprolactone).
- d. Polyesters produced by microorganisms; e.g., poly (hydroxyal kanoate).

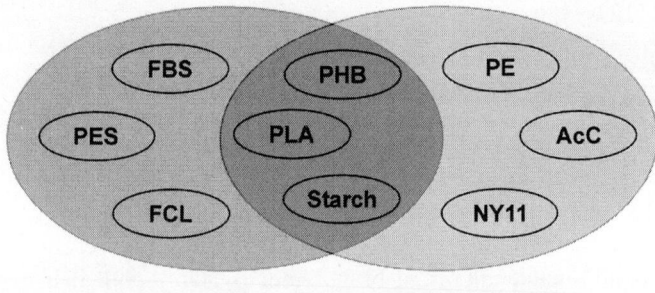

Degradable Polymers   **Polymers derived from renewable resources**

Figure 2.7. Biodegradable and sustainable fibres

2.9.3.1   Sustainable technologies and practices

Sustainable technologies and practices includes the following:

Green dyes

- Extraction from plants
- Extraction from arthropods and marine invertebrates (e.g., sea urchins and starfish)
- Extraction from algae (e.g., blue–green algae)
- Production from bacteria and fungi.

Processes

- Cold pad batch preparation and dyeing
- Continuous processing of knits
- 1&2 stage vs. 3-stage preparation of wovens
- Combined scour & bleach for knit and yarn
- Foam dyeing, finishing and coating
- Pad/dry vs. pad/dry/pad/steam.

Chemicals and dyes

- Cat ionisation for salt-free dyeing
- Stable chemistries for 1 or 2-stage vs. 3-stage preparation
- High fixation dyeing with reduced salt
- Enzymatic desizing and scouring
- Size recovery and recycle
- Liquid indigo and sulphur dyes
- Pigment printing and dyeing
- Right First Time (RFT) dyeing.

Equipments

- Low liquor ratio jets with LR less than 8:1
- Low liquor ratio package dyeing with LR less than 6:1
- Filtration of process water for recycle
- Caustic recovery and re-use
- Insulated dyeing, drying and stenter machines
- Solar heating of water.

Systems, control and management

- Empowered environmental teams
- Automatic dyes and chemicals dispensing
- Advanced equipment and process control
- Various system approaches to reduce WEC.

Waste water treatment

- Physical, biological and activated carbon systems
- High technology filtration systems
- Recycle internal process water
- Waste water treatment sludge used/sold for fuel.

2.9.4   New range of eco-fabrics

Relying on polluting textile materials like cotton and polyester may become a thing of the past as a new range of eco-fabrics emerge, often made from materials that would otherwise go to waste. Some of these environmentally friendly fabrics are already in use, like those made of coconut husks, recycled plastic bottles, wood pulp and corn, while others are strange and futuristic, sourced from hagfish slime, fermented wine, spoiled milk and genetically engineered bacteria.

Figure 2.8. Fabric from fermented wine

A group of scientists at the University of Western Australia has produced fabric by letting microbes go to work on wine. The scientist culture bacteria called *Acetobacter* in vats of cheap red wine, and the bacteria ferments the alcohol into fibres that float just above the surface. These fibres can be extracted and fashioned into clothing (Figure 2.8).

2.9.5   Naoran, durable fabric made of wood pulp

This leather alternative is not only animal-friendly, it also eschews the chemicals required to create conventional faux leather. Naoran is a water-resistant textile derived from wood pulp and recycled polyester. It's soft, flexible, and tear- and water-resistant (Figure 2.9).

Figure 2.9. Apparel made from wood pulp

2.9.6   Hagfish slime thread

The slimy substance in the photo above is defensive goo attached to a hagfish, an eel-shaped bottom-dwelling animal of the deep seas that is the only known creature to have a skull, but no vertebral column. Scientists have discovered that proteins within this slime have mechanical properties rivalling those of spider silk, and can be woven into high-performance bio-materials (Figure 2.10).

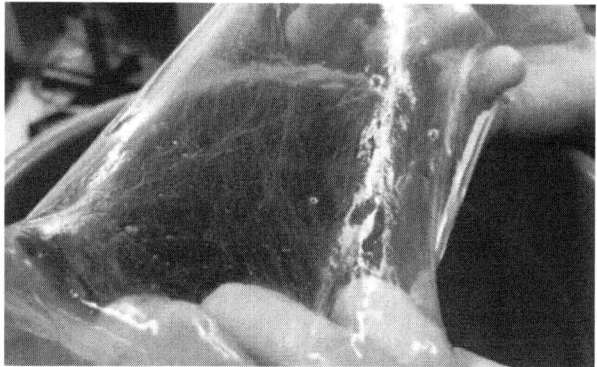

Figure 2.10. Hagfish slime thread

2.9.7   Spider silk made from metabolically engineered bacteria

Known for its tremendous strength—three times stronger than both steel and Kevlar, yet thinner than a human hair——spider silk is a highly valuable material for textiles. But farming and harvesting spider silk is a definite challenge. Instead, geneticists have found a way to chemically synthesize the silk gene and insert it into *E. coli* bacteria (Figure 2.11).

Figure 2.11. Spider silk web

2.9.8   Ingeo, fabric made from corn

Synthetic fibres are most often petroleum-based, but recycled fibres and those sourced from natural substances are on the rise. Ingeo, a fabric by nature works derived from fermented corn starches, can be spun into fibers for apparel and home textiles, and also used for bio-plastics (Figure 2.12).

Figure 2.12. Fabric made from corn

2.10   Silk-like fibre derived from spoiled milk

Few of us would willingly walk around wearing spoiled milk, but it might just become all the rage in the near future. A company called Qmilch makes fabric from protein found in soured "secondary milk" that's no longer suitable for human consumption, and would normally be thrown away. This zero-waste fabric requires no harmful chemicals to make, and uses less water in the production process than other milk-based fabrics.

2.10.1   New life polyester yarn made of recycled plastic bottles

New life is a polyester yarn made from 100% post-consumer recycled plastic bottles, which is processed by mechanical rather than chemical means. Made in Italy, the fabric is used in fashion, sportswear, underwear, medical garments and other clothes and furnishings. Georgio Armani used it to create a fashionable, eco-friendly gown for Livia Firth at the 2012 Golden Globe Awards (Figure 2.13).

Figure 2.13. Recycled fibres

2.10.2   Used coffee pods

Inspired by the resourcefulness of locals in Kerala, India, who repurpose waste in surprising ways, designer Rachel Rodwell discovered a material that wasn't living up to its potential—used coffee pods. Rodwell gathers pods from friends and family and smashes them with a meat tenderizer, reconfiguring them into geometric-inspired designs in colours that reflect India's cultural aesthetics.

2.10.3   Recycled newspaper yarn

Figure 2.14. Recycled paper yarn

Artist IvanoVitali tears recycled newsprint into strips and twists it into balls of yarn without the use of glue, dyes or silicone, crocheting them into textile art with custom-made wooden knitting needles and hooks as long as 8 feet. Recently, Vitalihas expanded into wearable art, achieving certain colours for dresses, jackets and even bikinis by painstakingly sorting his printed materials by colour (Figure 2.14).

2.10.4   Self-repairing textile

Once a protective garment like a raincoat or lab wear is ripped or torn, it's useless. But the total loss of these garments may become a thing of the past with the creation of "intelligent" fabric that can heal itself. Researchers at SINTEF added microcapsules containing a glue-like substance to the plastic polyurethane that is applied to modern rainwear, so that if the garment snags, the capsules release a sealant that fills in the gaps and hardens with contact to air and water.

2.10.5   Lab-grown biological textiles

How will biotechnology change the fashion industry? Designer Amy Condon believes that in the future, we'll be able to grow textiles like ethical issues in laboratories. Her series "Biological Atelier" imagines a workshop, circa 2082, where high-fashion garments are grown from cells. Considering the social and environmental impact of the textile industry, even the most unlikely sounding ideas deserve a good look—and while biological Atelier is intended for bespoke luxury garments, similar technology could possibly be used on a wider scale. When grown in a lab or made from waste materials rather than farmed, biological-based textiles could potentially replace fabrics made from unsustainable materials, like polyester—as long as they don't require too much chemical manipulation in the process (Figure 2.15).

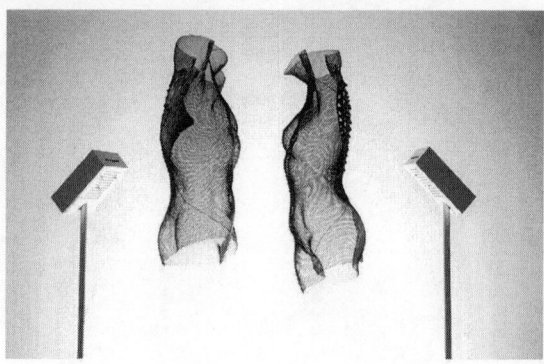

Figure 2.15. Biologic second skin, Tangible Media Group, MIT Media Lab.
Photo courtesy: © 2012 Tangible Media Group / MIT Media Lab

2.11   Degradation mechanisms

Hydrolytic degradation: Most hetero-chained polymers are susceptible to aqueous acid or base degradation referred to as hydrolysis. This susceptibility is due to a combination of the chemical reactivity of heteroatom sites and to the materials being at least wetted by the aqueous solution, allowing contact between the protons or hydroxyl ion to occur.

Photo-degradability: Synthetic polymers can be degraded by the absorption of ultra-violet light. This applies also to natural polymers, but these tend to be more rapidly degraded biologically, by the attack of micro-organisms.

Biodegradation: In the broadest sense, biodegradation is the biologically catalysed reduction in the size and complexity of a molecule (Figure 2.16). This breakdown is carried out by microorganisms which, because they are living entities, require suitable conditions such as optimum pH and temperature in the composting process.

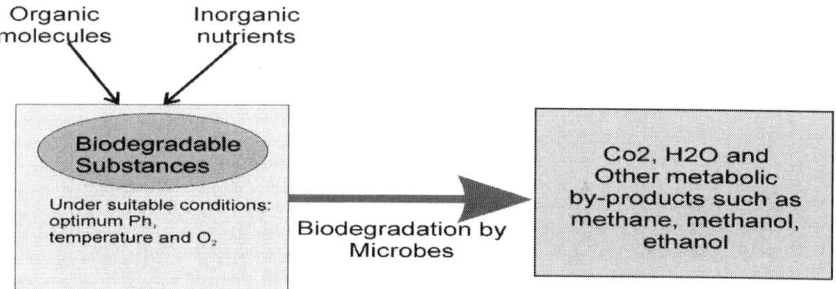

Figure 2.16. Degradation mechanisms

2.12   Sustainability issues in textile dyeing

Commercial dyeing can be described as a method for colouring a textile material in which a dye is applied to the substrate in as uniform a manner as possible to obtain an even shade (or level dyeing) with a performance and fastness appropriate to its final end-use. Textile dyeing involves the use of a number of different chemicals and auxiliaries to assist the dyeing process. Some auxiliaries (e.g., dispersing agents, buffers, dedusting agents) are already contained in the dyestuff formulation, but other auxiliary chemicals are also added during processing to aid the preparation, colouration and after washing processes. Since auxiliaries in general do not remain on the substrate after dyeing, they are ultimately found in the emissions from a dye house. Environmentally responsible dye application embraces the well-established "3R" principles of pollution prevention—i.e., reduce, re-use, recycle—and the most effective pollution prevention practice for textile wet processing is "right-first-time" dyeing. Corrective measures such as strip and re-dye are all chemical, energy and water intensive and add significantly to the pollution load (Table 2.3).

Table 2.3. Type of pollution associated with various coloration processes.

Fibre	Dye class	Type of pollution*
Cotton	Direct	1 – salt
		3 – unfixed dye (5–30%)
		5 – copper salts, cationic fixing agents
	Reactive	1 – salt, alkali
		3 – unfixed dye (10–40%)
	Vat	1 – alkali, oxidising agents
		2 – reducing agents
	Sulphur	1 – alkali, oxidising agents
		2 – reducing agents
		3 – unfixed dye (20–40%)
	Chrome	2 – organic acids
		5 – heavy metal salts
Wool	1:2 metal complex acid	2 – organic acids
		2 – organic acids
		3 – unfixed dye (5–20%)
Polyester	Disperse	2 – reducing agents, organic acids
		5 – carriers

*Pollution categories: 1, relatively harmless inorganic pollutants; 2, readily biodegradable, moderate–high biological oxygen demand (BOD); 3, dyes and polymers difficult to degrade; 4, difficult to biodegrade, moderate BOD; 5, unsuitable for conventional biological treatment, negligible BOD.

2.12.1   Factors for improving sustainability in dyeing and finishing

- Accurate colour communication
- Intelligent dye selection for product durability
- Intelligent dye selection for chemical compliance
- Intelligent process selection for improved resource efficiency
- Waste minimisation and pollution control.

Table 2.4. Percentage of unfixed dyes in dyeing of different fibrous materials.

Fibre	Dye type	Unfixed dye %
Wool and nylon	Acid dyes/reactive dyes for wool	7–20
	Pre-metallised dyes	2–7
	After chromes	1–2
Cotton and viscose	Azoic dyes	5–10
	Reactive dyes	20–50
	Direct dyes	5–20
	Pigment	1
	Vat dyes	5–20
	Sulphur dyes	30–40
Polyester	Disperse	8–20
Acrylic	Modified basic	2–3

2.12.2   Azo-free colorants

Azo-free colourants are dyes and pigments that are free of the nitrogen-based compound aromatic amines, also referred to as "Azos". These compounds are toxic and banned in the European Union (EU) due to their mutagenic, carcinogenic and often allergic properties. These dyes are not biodegradable (Table 2.4).

2.12.3   Biodegradable dyes

Biodegradable refers to dyes that do not require the use of inorganic salts, heavy metals and amines. They are substances that decompose readily and become absorbed by the environment.

2.12.4   Chrome-free tanning

Chrome-free tanning is the tanning of hides to create leather either through the use of oils or natural tannins instead of chromium salts. This tanning process is more time intensive than chrome tanning, but is better for the environment, as the chromium method uses chrome, a known carcinogen that can be absorbed through the skin and cause contamination of soil and waterways surrounding tanneries.

2.12.5　Fibre reactive dyes

Fibre reactive dyes are dyes used to colour cellulosic and protein fibres such as cotton, rayon and soy. The dyestuff bonds to the fibres through a chemical reaction and does not require the use of mordants. Therefore, direct dyes require less salts and heavy metals to be used to achieve optimal colouration and fixation than other commodity dyestuffs. When used correctly, this can reduce not only the salt and metal content of the effluent, but also the quantity of water used to remove excess dye and the amount of dye run-off.

2.12.6　Heavy metal free dyes

Heavy metal free refers to dyes that do not require the use of heavy metals to achieve the fixation of colours. Toxic heavy metals, such as chrome, copper and zinc, which are all known carcinogens, are commonly used as fixers in dyes. Although most heavy metals can be removed from the effluent through waste water treatment, this often does not occur.

2.12.7　Low-impact dyes

Low-impact refers to synthetic dyes that do not use substantial levels of heavy metals or toxic chemicals as fixers.

2.12.8　Natural dyes

Natural dyes are dyes that are created from bark, bugs, flowers, minerals, rust and other natural materials. Natural dyes allow small producers to retain their traditional dyeing methods and promote biodiversity. The disadvantage of these dyes is that their mordants are often heavy metals.

2.12.9　Non-toxic semi-aniline dyes

Non-toxic semi-aniline dyes are non-toxic transparent dyes used to dye leather. These dyes are derived from coal tar and fully penetrate the leather while preserving the appearance of natural grains and markings.

2.13　Important environmental negative effects related to textile wet processing

- Chemical intensive wet processing—scouring, bleaching, mercerising, dyeing, printing, etc.

- Heavy metals—iron, copper, lead, etc. found in dyestuffs auxiliaries, binders, etc.
- Residual dyestuffs chemicals in water—due to poor fixation of colours.
- PVC and phthalates—used in plastisol printing paste.
- Formaldehyde—found in dispersing agents, printing paste and colorant fixatives.
- Dye effluent—waste water is used.

2.13.1   Chlorine-free bleaching

Chlorine-free bleaching is the use of hydrogen peroxide to whiten fabrics. Hydrogen peroxide naturally degrades into oxygen and water, leaving no harmful chemical residue on the cloth or in the effluent. It is sometimes referred to as Green Bleach.

2.13.2   Dry-heat fixation

Dry-heat fixation is a method of fixing reactive dyes printed through the ink-jet method. The dyed/printed fabric is passed through hot iron plates in lieu of steam. This method conserves water and energy by using an alternative to steam fixing as well as the ink-jet printing method.

2.13.3   Dye bath reuse

Dye bath reuse is the practice of recycling the water used in dye baths for subsequent baths. The water conserved through the bath's reuse is substantial, as anywhere from 10–50% of dye from one bath does not fix to the fabric.

2.13.4   Eco-bleach

Eco-bleach is the use of natural phosphates and silicates in cow dung combined with sunlight to achieve whitening of natural fabrics. This is the most eco-friendly form of bleaching.

2.13.5   Ink-jet printing

Ink-jet printing is a method of applying pigment and dyes to cloth using an ink-jet printer. It is considered the most eco-friendly and efficient method of printing due to its lower water wastage and energy consumption compared with other commercial printing methods.

2.13.6   Vegetable tanning

Vegetable tanning refers to the use of natural tannins to create usable leather from hides. Natural tannins are present in bark, wood, leaves and fruits of chestnut, oak and hemlock trees. This process is time intensive, as it can take-up to 3 weeks for the tannins to fully penetrate a hide. From an ecological perspective, vegetable tanning is preferable, however, the leather produced is not stable in water as it shrivels and becomes brittle.

2.13.7   Waste water recycling

Waste water recycling is the use of tertiary treated waste water in the dye baths and/or for irrigation purposes. This water is suitable for human contact but is not potable. Its use reduces the strain on potable water supplies particularly in arid climates and is an effective way to reuse this valuable resource.

2.14   Green chemistry

Anastas and Warner defined the term "green chemistry" as the design of chemical products and processes that reduce or eliminate the use or generation of hazardous substances [P. Anastas and J. C. Warner, Green Chemistry: Theory and Practice, Oxford University Press, New York, 1998].

2.14.1   Principles of green chemistry

- **Waste management:** Elimination or minimisation of waste;
- **Atom economy:** No or lower wastage of atoms;
- **Catalysis:** Catalysts are proffered to stoichiometric reagents;
- **Direct reactions:** Use of minimum or fewer reaction steps; derivates or intermediate steps use additional reagents and have the potential to generate more waste;
- **Safer reactions:** Synthetic methodologies should be designed to use and generate substances that possess little or no toxicity to human health and the environment;
- **Renewable raw materials:** Use of renewable and non-depleting feed stocks;
- **Safer production:** Preservation of the efficacy of functioning while reducing toxicity;
- **Biodegradability:** Use of easily and harmlessly degradable chemicals with no accumulation in the environment;
- **Green auxiliaries:** Use of auxiliary substances (e.g., solvents, separation agents) should be avoided wherever possible and where absolutely necessary, should be preferably be innocuous;

- **Energy economy:** Saving the energy should be achieved preferably by using reactions that take place under ambient temperature and pressure conditions;
- **Safer by-products:** Real-time monitoring and control and reuse of by-products;
- **Hazard control:** Avoidance of hazardous chemicals to minimise the chance of explosions, fire and harmful releases.

2.14.2   Non-eco-friendly substances

The terms "environmentally friendly", "eco-friendly", "nature friendly" and "green" are used to refer to goods and services, laws, guidelines and policies claim to inflict minimal or no harm to the environment. "Green" is a very subjective term that could be interpreted in different ways, but whatever the definition, becoming green is important because it means being committed to protect the people and the planet. Green or eco-friendly goods, services and practices assure the use of environmentally—friendly materials, free from harmful chemicals, compounds or energy waste, which do not deplete the environment during pollution and transportation, whereas non-eco-friendly substances such as non-biodegradable organic materials, and hazardous substances may do harm to the environment.

The main non-eco-friendly substances are as follows:

- **Non-biodegradable organic materials:** Substances that are not broken down by microorganisms, have no biochemical oxygen demand (BOD), and have an oxygen demand only if it is a chemical reducing agent;
- **Hazardous chemicals:** Those who are a physical hazard or a health hazard;
- **Toxic metals and heavy metals:** Bjerrun defined "heavy metals" those with elemental densities above 7 g/cm^3; over the years this definition has been modified by various authors;
- **Toxic volatile organic compounds (VOCs):** Are organic chemicals that have a high vapour pressure under normal atmospheric conditions;
- **Hazardous substances in transit:** Due to unprecedented growth of chemical industries in the developing countries, the proportion of the total freight traffic involved in the transport of hazardous chemicals is increasing at a rapid rate. Of the carriers transporting hazardous goods, the majority carry flammable petroleum products including kerosene, petrol, liquefied petroleum gas and naphtha (Table 2.5).

Table 2.5. Health hazards associated with some heavy metals and metalloids used in the textile industry.

Metal/metalloid	Associated health hazards
Lead (Pb)	Damage to the brain, nervous system and kidneys (causes in mild cases insomnia, restlessness, loss of appetite and gastrointestinal problems)
Mercury (Hg)	Damage to the brain
Cadmium (Cd)	Disorders of the respiratory system, kidneys and lungs
Chromium (Cr)	Skin and respiratory disorders, ulceration of skin and cancer of the respiratory tract on inhalation
Arsenic (As)	Skin cancer, hyperpigmentation, kurtosis and black foot disease

[Source: D. A. Phillips, Chemistry and biochemistry of trace metals in biological systems, in Eeffect of Hheavy Mmetal Ppollution on Pplants, N.W. Lepp, ed. Applied Science Publisher, Barking, 1981.]

2.14.3   Characteristics of green chemicals

1. Prepared from renewable or readily-available resources by environmentally-friendly processes;
2. Low tendency to undergo sudden, violent, unpredictable reactions such as explosions;
3. Non-flammable or poorly flammable;
4. Low-toxicity and absence of toxic constituents, particularly heavy metals;
5. Biodegradable;
6. Low-tendency to undergo bio-accumulation in food chains in the environment.

2.15   Dyestuffs and finishing—chemistry is essential and ecological

Chemistry plays a significant role in textile finishing. Natural dyestuffs have been replaced little by little by synthetic dyestuffs. In the past, dyestuffs like indigo were extracted from plants. In 1878 the first production of synthetic dyestuff succeeded. Since 1897 synthetically produced indigo has been commercially marketed and has almost completely replaced the indigo production out of vegetable raw materials. In terms of economy and ecology the synthetic process turned out to be significantly superior to the natural process.

In addition, the auxiliaries for finishing are mainly on synthetic chemical basis. The reason for this development is the predominance of synthetic products with regard to dye fastness and levelness of finishing. Petroleum is mostly used for the production of chemicals. As this is not a regenerative raw material, a careful handling with these chemicals and auxiliaries is necessary.

One may reasonably expect that in an environment shaped by competition the costs for the production factors lead the dye houses to a possibly careful handling with them. However, by taking into account this criterion, no statement can be made regarding the environmental friendliness and sustainability (Table 2.6).

2.15.1 Regenerative raw materials replace fossil ones: the example of chitosan

Subject to certain conditions it makes sense to do without a petroleum based production of substances, but to use natural regenerative raw materials. The example of indigo shows that regenerative raw materials cannot be preferred generally. Chitosan is a regenerative raw material which could replace traditional chemistry—yarns processed on modern weaving machines have to be treated with synthetic sizing agents. These agents protect the warp yarns from high rubbing forces by harness frames and neighbouring threads in the automatic weaving process. However, they are not biologically degradable or only under certain conditions. Conventional sizing agents are based on starch, CMS, CMC, PVA, PET and acrylates. The latter ones are mineral oil based and show particularly good effects. Sizing agents based on chitosan come into consideration as an environmentally friendly alternative. Chitosan is extracted from chitin which is besides cellulose the most abundant polysaccharide on earth. Chitosan can be produced by separation of acetyl groups. This polymer is non-toxic and excellently biologically degradable.

Table 2.6. Eco-friendly alternative chemicals for textile wet processing.

S. No.	Purpose	Chemical	Alternative
1	Sizing	Starch	Water-soluble polyvinyl alcohol
2	Desizing	Hydrochloric acid	Amylases
3	Scouring of cotton	Sodium hydroxide	Pectinases
4	Bleaching	Hypochlorites	Hydrogen peroxide
5	Oxidation of vat and sulphur dyes	Potassium dichromate	Hydrogen peroxide, sodium perborate
6	Thickener	Kerosene	Water-based polyacrylate co-polymers

7	Hydrotropic agent	Urea	Dicyanamide (partially)
8	Water repellents	C8 fluorocarbons	C6 fluorocarbons
9	Crease recovery chemicals	Formaldehyde-based resin	Polycarboxylic acids
10	Wetting agents and detergents	Alkyl phenol ethoxylates	Fatty alcohol phenol ethoxylates
11	Neutralisation agent	Acetic acid	Formic acid
12	Peroxide killer	Sodium thiosulphate	Catalases
13	Mercerisation	Sodium hydroxide	Liquid ammonia
14	Reducing agents	Sodium sulphide	Glucose, acetyl acetone, thiourea dioxide
15	Dyeing	Powder form of sulphur dyes	Pre-reduced dyes
16	Flame retardant	Bromated diphenyl ethers	Combination of inorganic salts and phosphates
17	Shrink proofing	Chlorination	Plasma treatment

2.16   References

1. Md Mashiur Rahman Khan and Md Mazedul Islam, Materials and manufacturing environmental sustainability evaluation of apparel product: knitted T-shirt case study, Textiles and Clothing Sustainability, 2015, 1–2.
2. Richard Blackburn, Sustainable Aapparel, Woodhead Publishing Series in Textiles , 2015, 171, 2–28.
3. Subramanian Senthil Kannan Muthu, Sustainable apparel production, CRC Publishing, 2014, 13, 516–525.
4. Aparna Sharma, Eco-Friendly Textiles: A Boost to Sustainability, TMU, 2–8.
5. Ayatullah Hosne Asif, A.K.M., An Overview of sustainability on apparel manufacturing industry in Bangladesh, Science Journal of Energy Engineering, Vol. 5, No. 1, 2017, 1–3.
6. Sujata Saxena, A.S.M. Raja and A. Arputharaj, Challenges in Sustainable Wet Processing of Textiles, Textiles and Clothing Sustainability, 2017, 9–20.
7. Muhammad Adnan Ali, Muhammad Imran Sarwar, Sustainable and Environmental Friendly Fibers in Textile Fashion, University of Borås, 2010, 10–22.
8. Friedrich Petry, Environmental protection and sustainability in the textile industry, Textile Finishing, 2008, 1–3.
9. Antonela Curteza,, Sustainable Textiles, "Gheorghe Asachi" Technical University of Iasi, 2–83.
10. R. S. Blackburn, Sustainable textiles Life cycle and environmental impact, Woodhead Publishing, 2009, 98, 69–188.
11. Boström, M.; Micheletti, M. Introducing the sustainability challenge of textiles and clothing, 2016, 39, 367–375.

12. Allwood, J.M.; Laursen, S.E.; Russell, S.N.; de Rodriguez, C.M.; Bocken, N.M.P. An approach to scenario analysis of the sustainability of an industrial sector applied to clothing and textiles, 2008, 16, 1234–1246.

13. Guo, Z.; Liu, H.; Zhang, D.; Yang, J. Green supplier evaluation and selection in apparel manufacturing using a fuzzy multi-criteria decision-making approach. Sustainability 2017, 9, 650.

14. C. A. Rusinko, "Green manufacturing: An evaluation of environmentally manufacturing practices and their impact on competitive outcomes," IEEE Transactions on Engineering Management, vol. 54, no. 3, 2007, 445–454.

15. Fletcher, K., Sustainable Fashion and Textiles: Design Journeys, Published by Earthscan, London, 2008, 4–8.

3
Eco-hazards in manufacturing of apparels

S. Manjula M.Tech., P. G. Dip. (CD&BC)., NET.,
Assistant Professor, Department of Costume Design and Fashion, Kongu Arts and Science College (Autonomous), Nanjanapuram, Erode-638107.
Email id: manjulalokgu@gmail.com

Abstract: Ecological considerations have become important factors in marketing of textile goods. Manufacturers must be aware of the growing ecological concerns starting right from the stages of cultivation and production of fibres upto packaging. The environmental hazards faced due to the usage of pesticides and its alternatives are discussed. The problems associated with the textile processing industry are complex in nature. The health hazards encountered as a result of the application of banned dyes and chemicals are elaborated and their permissible limits in textiles are highlighted.

Keywords: pesticides, allergic dyes, toxic chemicals, dermatitis, hazards, eco norms.

3.1   Introduction

Textiles make up a very broad category of products and are used in a way that consumers, including children, are directly or indirectly exposed to their chemical content. Chemicals in textiles can have adverse effects by directly affecting health, such as causing allergic reactions. But they can also adversely affect the environment, for example, by long-term effects from persistent or bio-accumulating substances.

Textile industry contributes 30% of India's export. It produces over 400 million meters of cloth and around 1000 million kg of yarn per annum. Textile sector is labour intensive and nearly a million of workers are associated in various unit operations. Textile wet processing activity contributes about 70% of pollution in textile industry.[1] It is estimated that there are around 12,500 textile processing units wherein the requirement of water ranges from 10 litres with an average of 100 litres per kg. Right from cotton cultivation and manufacture of fibres, spinning, weaving, processing and finishing, more than 14,000 dyes and chemicals are used and a significant quantity of these goes in the solid, liquid and air wastes, thereby imparting pollution of air, land and surface water. To deal with the problems posed by hazardous chemicals in textiles, there is a need for regulation of chemical content in textile products.[2]

3.2   Pesticides in cotton cultivation

Cotton is a highly pest-prone crop. Large quantities of most acutely toxic pesticides are used in conventional cotton cultivation. The proportion of global pesticide consumption for cotton has averaged 11%. These chemicals account for more than 50% of the total cost of cotton production in much of the world. In addition to insecticides, significant amount of herbicides, fungicides, defoliants as well as synthetic fertilizers are used in cotton cultivation. Nitrogenous and phosphatic fertilizers are widely used in cultivation.[3] This residual nitrates and phosphates present in agricultural drainage cause water pollution. Ground water gets polluted due to seepage from the surface. The additions of chemicals reduce the productive capacity of the soil. Toxic insecticides kill useful soil bacteria which are essential for plant growth. Chlorinated pesticides convert to dioxins which are highly toxic. Some pesticides are also carcinogenic. In various stages, the pesticides are harmful to the environment through soil, air and water pollution.

Figure 3.1. Organic cotton fibre label

3.3   Organic cotton

For cultivation of organic cotton, chemical fertilizers and pesticides are not used. In order to remove the residual pesticides in the soil, crops are to be cultivated for three seasons without the use of any chemical fertilizers and pesticides. Organic production systems replenish and maintain soil fertility, reduce the use of toxic and persistent pesticides and fertilizers, and build biologically diverse agriculture.[4]

The same rule of organic cultivation is applied to flax, wool, silk, jute, ramie, hemp, etc. Farmers should be advised to use only permissible pesticides such as neem oil, permethin, cypermethin, etc. Organic cotton cultivation helps in improving soil fertility and in decreasing pollution. Organic cotton fibre is used in everything from personal care items (sanitary products, make-up removal pads, cotton puffs, ear swabs), to fabrics, home furnishings (towels, bathrobes, sheets, blankets, bedding, beds), children's products (toys, diapers), and clothes of all kinds and styles (whether for lounging, sports or the workplace). In addition, organic cottonseed is used for animal feed, and organic cottonseed oil is used in a variety of food products, including cookies and chips.

3.4   BT cotton

BT cotton, genetically engineered (transgenic) cotton, was heralded for its environmental and human health benefits and as a step towards sustainable agriculture since, farmers could significantly reduce insecticide use. To create cotton with built-in protection against insects, genetic engineers spliced a BT toxin gene into cotton (Fig. 3.2). The new gene that enabled the transgenic cotton to produce insecticidal toxin throughout the plant was obtained from a soil bacterium, *Bacillus thuringiensis* (Bt), an organism well known to many organic and sustainable growers who have used Bt in sprays to control insects. Bt gene is put into cotton to protect against three pests—tobacco budworm, cotton bollworm and pink bollworm. However, the Bt cotton is not effective against a number of other pests, including the boll weevil and whitefly.[5]

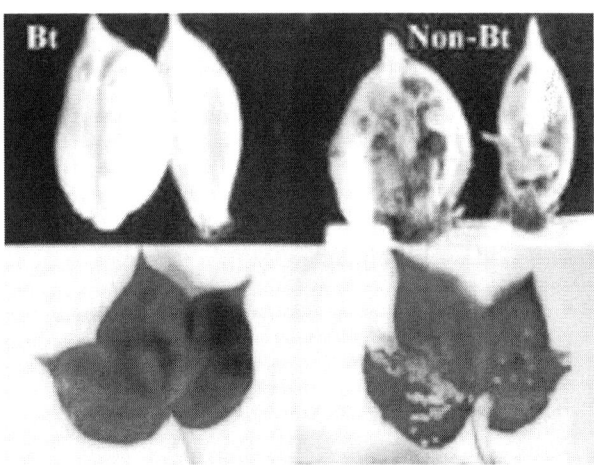

Figure 3.2. Comparison of Bt cotton and Non-Bt cotton

3.5   Health hazards

The consumer of non-organic textile products may suffer from negative health impacts caused by hazardous substances primarily via dermal exposure, indirect dermal and oral exposure via indoor dust, and indirect oral exposure through food. Children are particularly vulnerable and may, due to their proneness to put things in their mouths, be directly exposed to hazardous chemicals in textiles.[6]

Figure 3.3. A brand of chemical-free clothing

3.6   Heavy metal content

Heavy metals are constituents of some dyes and pigments. They can also be found in natural fibres due to absorption by plants through soil. Metals may also be introduced into textiles through dyeing and finishing processes. After chemical processing of textiles, waste water contains many impurities as well as chemicals. Each year the global textile industry discharges 40,000–50,000 tons of dye into our rivers rivers (Fig. 3.4), and more than 200,000 tons of salt.

Effluent from textile dyeing process contains heavy metals. These metals are toxic as their ions or compounds are soluble in water and may be readily absorbed into living organisms. These heavy metals which have transferred to the environment are highly toxic and can bio-accumulate in the human body, aquatic life, natural water-bodies and also possibly trapped in the soil. Once absorbed by humans, heavy metals tend to accumulate in internal organs such as the liver or kidney. The effects on health can be tremendous when high levels of accumulation are reached. For example, high levels of lead can seriously affect the nervous system. After absorption, even in small amounts these metals bind to structural proteins, enzymes and nucleic acids causing health effects. The toxicity of metal pollution is slow and long lasting as these metal ions are non-degradable. Among the different heavy metals iron, copper, aluminium and tin are considered more safe compared to lead, chromium, cadmium and mercury.[7]

Heavy metals very often refer to Antimony (Sb), Arsenic (As), Lead (Pb), Cadmium (Cd), Mercury (Hg), Copper (Cu), Chromium (Cr), Cobalt (Co), Nickel (Ni). Both cadmium and lead are classified as carcinogens. There are no legal limits for heavy metal contents in textiles. However, eco-labels and buyers have adopted limits of drinking water for textile end products. Maximum permissible limits of heavy metals in drinking water according to Indian standards are less than 1 ppm in almost all the cases.

Figure 3.4. Effluents discharged from textile dyeing industry

Prolonged exposure to heavy metals may cause health problems such as kidney failure, emphysema, allergies and even cancer. Copper causes irritation of mucous membrane. Iron when contact with skin and eyes causes severe burns. Manganese causes symptoms of Parkinson's disease. Aluminium is a probable cause for Alzheimer's & Parkinson's disease. Zinc causes gastrointestinal problems and damage to kidneys. Selenium causes nausea, diarrhoea, fatigue and hair loss. Lead affects the central nervous system. Mercury affects central nervous system and the areas associated with visual and auditory functions. Cadmium causes kidney failure. Nickel induces irritation of skin and eyes and dermatitis. Long continuous exposure to chromium may lead to liver and kidney disease and even cancer.[8]

3.7   Pentachlorophenol

Generally chlorinated phenols have strong biological effects. Of all chlorinated phenols, pentachlorophenol (PCP) seems to have been produced and used in large quantities. Pentachlorophenol is used as a preservative of size and gums. It is used as dispersant in latex based finishes. PCP and its salts are used as algaecides, fungicides, bactericides, herbicides, insecticides and mollucides. It is used widely in preservation of wood and as preservative in pigment emulsions. Adhesives, glues, vegetable and mutton tallow may also have PCP additives.

It has been found that it is toxic to animals and humans. Developmental and reproductive effects, liver and kidney pathology were noted in animal studies. PCP is highly toxic to aquatic organisms. Invertebrates and fish are adversely affected by concentrations of PCP below 1 mg/l. Algae are very sensitive. PCP contains several highly toxic dioxins which have shown carcinogenic effects in experimental animals. Acute exposure to PCP can cause elevated temperature, profuse sweating, dehydration, loss of appetite, decreased body weight, nausea, uncoordinated movement and coma. Exposure may be reduced by providing protective clothing. Where dermal contact is expected, wear gloves. Spray applicators should wear protective clothing during spraying. Automated processes and use of closed systems reduce work exposures.[9]

3.8   Formaldehyde

At room temperature, formaldehyde is a colourless, flammable gas with a distinct pungent smell. It is also known as methanol, methylene oxide and oxomethane. It is present in nature in small quantities. For example, human blood has traces of formaldehyde and so do apples. However, in large quantities it may be a skin irritant.[10]

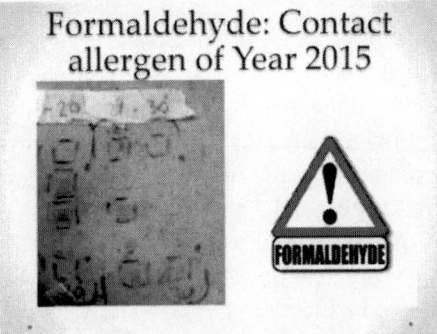

Figure 3.5. Formaldehyde and its health hazard

3.8.1   Sources of formaldehyde in textiles

It is used as a mordant (dye fixing agent) and fixers and binders in pigment printed fabrics. It is used in anti-shrinking treatments and in resin finishes for wrinkle/crease-free properties. And it is used in tanned leather products and cosmetics.

3.8.2   Hazards

Low levels of formaldehyde can cause irritation of the eyes, nose, throat and skin. People with asthma may be more sensitive to the effects of inhaled formaldehyde. In animal studies, rats exposed to high levels of formaldehyde in air developed nose cancer. It is an allergenic (Fig.3.4) and a skin irritant in large quantities. The Department of Health and Human Sciences has determined that formaldehyde may reasonably be anticipated to be a carcinogen.[11]

3.8.3   Limits

The occupational safety and health administration has set a permissible exposure limit for formaldehyde of 0.75 ppm for an 8-hour workday. The short-term exposure limit of 15 minutes is 2 ppm.

3.9   Nickel

Nickel presence is known to cause contact dermatitis in approximately 10% of the European population. Therefore, a regulation restricting the amount of nickel that is released by metal accessories was necessary. Use of nickel was restricted in metallic accessories by many countries in Europe, due to the potential to cause skin allergies allergies (Fig. 3.6).[12]

Figure 3.6. Labelling terms used for safe jewelry

3.9.1   Reasons for nickel sensitivity

Ear piercing is an important risk factor for nickel sensitivity. Traditionally higher in females, its occurrence may change with men having pierced ears today and both sexes increasing the piercing of other body parts. Ear piercing may cause increased reaction to other metals such as gold and mercury.[13]

3.9.2   Nickel exposure

Examples of exposure to nickel include clothing snaps and buttons, hooks and garter snaps, rivets and zippers, identification tags, watch straps, blue jeans buttons, scissors and knitting needles, jewellery, thimbles, coins, keys and buckles.

3.9.3   Nickel sensitisation

Nickel sensitisation is aggravated in body locations when sweat leaches the nickel out of metal. Nickel may produce either allergic contact dermatitis or urticaria. With contact urticaria from nickel in necklaces not made of sterling silver or a high percent of gold, the neck begins to itch almost immediately and turns red.

3.9.4   Norms

- Nickel (general) = 0.02–10 mg/kg
- Nickel (baby clothing) = 0.02–1 mg/kg
- Deliveries containing nickel should be clearly marked, when the content is higher than 0.5 mg
- Consumer articles containing nickel and coming into contact with skin and releasing nickel must be provided with the declaration— "THIS PRODUCT CONTAINS NICKEL".

3.10   Sensitizing and allergic dyes

The main functions of clothes are protecting from environmental injuries and helping to regulate skin temperature and moisture. Natural and synthetic fabrics used in the manufacture of clothing cause almost no skin problems. Clothing dermatitis is generally attributed to chemicals and dyes added to these fibres during their manufacture and assembly into garments. In particular, responsible agents are represented by finishes, dyes, metals, rubber and glues. Also optical whitener, biocides, flame retardants and other agents are occasionally recognised as causative substance. Any location where the clothing is held more tightly against the skin is a likely spot for textile dermatitis. Sometimes, other sites not directly exposed to sensitising substances, as face and hands, can be involved as well.

Not all garment related allergy trouble can be attributed only to the dyestuffs used but also many other factors, however attention should be paid as some dyestuffs do contain allergenic (sensitising) properties. Parameters influencing the risk of sensitisation in textile dyes include not only the allergenicity of dye molecule but also the fastness of the dye i.e., how well it is bound to the fabric and the percutaneous absorption. An allergic skin reaction requires re-exposure or continuous exposure to the allergen, the substance causing the allergy. The first exposure "sensitises" the person and the later exposure "elicits" the reaction. The hypersensitivity to an allergen can be immediate (Type I)—a sometimes hazardous response that occurs within minutes to an hour of contact with the allergen—or delayed (Type IV), appearing at 24–72 hours.

3.10.1   Disperse dyes

Textile dyes are causes of acute dermatitis with rapid onset. Among textile dyes, disperse dyes are considered the most responsible. Literature data report their prevalence of sensitisation ranging between 3.1% and 5.2%. In particular, disperse blue dyes were recently selected as "contact allergens of the year" for 2000. Patch tests with disperse dyes performed on children with suspected ACD and/or atopic dermatitis have shown a positivity of 4, 6%—the most common sensitizer was Disperse Yellow 3. Both animal tests and human patch test studies have shown significant potential for sensitisation to disperse blue 35, 106 and 124. The above dyes have been reported to cause an ACD to a variety of garments which include underwear, blouses, pants, swimsuits, etc. There is evidence that atleast 15 disperse dyes are contact allergens particularly those applied to garments made of synthetic fibre and worn skin-tight. In the 70s this phenomenon was called as *"stockings dye allergy"* and in 90s as *"leggings allergy"*.

3.10.2   Basic and acid dyes

The basic dyes are the next most common allergens. They are mainly used to dye wool, silk, cotton, cellulosics and polyacrylonitriles. Basic Red 46, Basic Brown 1, Basic Black 1, Brilliant Green and Turquoise have been reported to cause textile dermatitis.

Acid dyes, also indicated as textile allergens are used to colour wool, other protein fibres and some manmade fibres (nylon). They include monobasic, diabolic, triphenylmethane and anthraquinone compounds. Acid Yellow 23, Supramine Yellow and Red and Acid Violet 17 belong to this class.

3.10.3   Direct dyes

Direct dyes are directly applied on fibres, most often wool, cotton, flax and leather. Water soluble direct dyeing agents are bound to the fibres by depositing in cavities. Binding is not very strong which means that the colour fastness is only moderate. However these agents are characterized by a low absorption through the skin. Direct Black 38, a triazoic compound dye, has been reported to be an allergen.

3.10.4   Vat dyes

Vat dyes are water-insoluble dyes applied in a reduced soluble form and then re-oxidized to the original insoluble form once absorbed into the fibre. They are used for cellulosic and some wool principally. Anthraquinone or indigoids are the chemical groups used. Vat dyes are relatively hypoallergenic although Vat Green 1, an anthraquinone derivative, has been reported to cause five cases of contact dermatitis from blue uniforms in nurses.

3.10.5   Reactive dyes

With an azo or anthraquinone structure, connected to a reactive group, capable of linking through covalent bonds to amine or sulfhydryl groups, these dyes are used for colouring natural fibres such as cotton, silk and wool and are also applied for polyamides. Asthma and contact dermatitis have been described after occupational exposure to reactive dyes. A study conducted on 1813 subjects patch tested with 12 reactive dyes has revealed 18 patients sensitised to these dyes. Most of them showed dermatitis localised to the trunk, upper limbs and/or hands; only one patient, occupationally exposed, presented dermatitis localised to the face. The dyes most frequently responsible for positive patch tests were Red Cibacron CR (Reactive Red 238) and Violet Remazol 5R (Reactive Violet 5). A further study performed with 5 other reactive dyes patch tested on 312 patients showed no positive allergic or irritant reactions.

3.10.6   Azo dyes

Regulations for azo dyestuffs are actually for certain azo dyestuffs that produce amine classified as carcinogenic due to reduction decomposition (Fig. 3.7). In the present context specific azo dyes releasing any of the 20 harmful amines has been banned as per the German Legislation. Azo dyes are manufactured from aromatic amines. It is important to note that some of them can split-off carcinogenic amines, such as benzidine, which may be absorbed through skin and the respiratory and intestinal tract. According to German Legislation, *"No articles of dress (textiles, leather, shoes) and bed linen can be put in trade, if these have been coloured with azo dyes that can release one of the twenty named amines"*.

Figure 3.7. International legislation on the prohibition of certain azo dyestuffs

3.11   Prominent eco-norms in textiles

- A ban on the use of PCP.
- In case of formaldehyde content above 1500 mg/kg in textiles, these must be marked by the declaration "contains formaldehyde".
- All supplies are expected to be free of carcinogenic substances and from acutely toxic (less than 200 mg/kg) dye and supplementary material, including organic chlorine and fire resistant chemicals.
- Dyes containing benzidine are to be avoided as they are known to produce toxicity.
- No halogenous dyestuffs containing bromide, chloride and fluorine besides urea should be used.
- Deliveries containing nickel should be clearly marked when the content is higher than 0.5 mg. Consumer articles containing nickel and coming into contact with skin and releasing nickel must be

provided with the following declaration: "This product contains nickel".

- Silks should not contain any heavy metal salts.
- While storing wool, only permitted insecticides are to be used.
- The accessories and trimmings used for garments must be eco-friendly: buttons be of natural materials, even coat hangers must be made of polystyrene according to DIN 6120.
- The industry may have to adopt a cradle to grave approach for the manufacture of eco-friendly textiles.

Figure 3.8. Hazardous chemicals and colorants used in apparels

3.12   Conclusion

Large quantities of chemicals are used in the manufacture of textiles. Some of these chemicals are harmful to human health and the environment and, for example, cause allergic reactions or bio-accumulating. Textiles make up a very broad category of products and are used in a way by which consumers, including children, are directly or indirectly exposed to the products' chemical content. It is difficult to know exactly which hazardous chemicals are present in textile products since the supply chains are long, complex and global. Today there is no unified legislation at the EU level covering the wide range of hazardous chemicals that may be present in textile products. There are however, a number of voluntary labels and restrictions lists used by the industry to limit chemical content in textiles. These voluntary efforts are not harmonised. In order to obtain a more cohesive handling of chemicals in textiles, there is a need for regulation at the EU level.

3.13   References

1. Sanjeev, R.S. (2006), Heavy metals, Colourage, 53(9–12): 66.
2. Kebschull, D. The challenge of eco-friendly textiles exports to Germany and Europe: Eco friendly textiles – Challenges to the textile industry, Textile Committee Publication. 1991, 28–39
3. http://infohouse.p2ric.org/ref/31/30335.pdf. Jan 1990
4. http://www.professional-laboratory.com/product_44.html. Professional Testing & Consulting Ltd. 2010
5. https://ota.com/sites/default/files/indexed_files/Organic-Cotton-Facts.pdf. May 2, 2017
6. Mathur, N., Bhatnagar, P., Bakre, P., (2005), Assessing mutagenicity of textile dyes from Pali (Rajasthan) using AMES bioassay. Applied Ecology and Environmental Research, 4: 111–18.
7. http://www.mindfully.org/GE/Another-Magic-Bullet.htm. Jan 16, 2013
8. Hatch, K.L. and Maibach, H.I., (2000), Textile dye allergic contact dermatitis prevalence. Contact Dermatitis, 42: 187–195.
9. Manzini, B.M., Motolese, A., Conti, A., Ferdani, G., Seidenari S., (1996), Sensitization to reactive textile dyes in patients with contact dermatitis. PubMed, -03-01.
10. http://nursinglink.monster.com/training/articles/299-skin-sensitization---contact-dermatitis-and-contact-urticaria. Jul 2017
11. http://pmep.cce.cornell.edu/profiles/extoxnet/metiram-propoxur/pentachlorophenol-ext.html. April 2018
12. Nadiger, G.S., (2001), Azo ban, econorms and testing. Indian Journal of Fibre and Textile Research, 26: 55–60.

4

Dyeing industry wastewater using reverse osmosis treatment practiced in Perundurai common effluent treatment plant—Part I

Dr. M. Ramesh Kumar and Dr. T. Saravana Kumar

Associate Professor, Department of Fashion Technology, Sona College of Technology, Salem – 636 005 Tamilnadu, India.
Email: rameshkumartex@gmail.com

Abstract: Textile industry plays an important role in Indian economy in multiple ways and it also causes major environmental impact. This super power continuously threatens the livelihood by discharging the effluent into nearby water body. On the other hand, water consumption for the plant is high. The two major process involved in dyeing process are preparation and colouration process. The preparation process includes singeing, desizing, scouring, mercerisation and bleaching; colouration process includes dyeing and printing process. The textile industry consumes a vast quantity of water and generates an equally vast quantity of wastewater. Textile wastewater is also known to exhibit strong colour, a large amount of suspended solids, non-linear behaviour of pH and high COD concentration. In this junction, a proper treatment is essential for safe and healthy atmosphere. The present work focus textile industries in which effluents are collected in two stages viz., wash water and dye bath effluents. The primary treatment which is based on the characteristics, treatment plant was developed. This is followed by the secondary treatment such as biological oxidation through tertiary clarifier, dual media filter, ultrafiltration (UF) and reverse osmosis (RO) system. These emphasised to recycle the water. RO permeate is 90% of water reused for processes. The rejects about 10% of the inlet volume is subjected to RO which are sent to evaporators. Dye bath water after treatment, the reject of about 20–25% is sent to multi effect evaporator system. This study was carried out in common effluent treatment plant (CETP), Perundurai, SIPCOT, Erode district.

Keywords: effluent, ultrafiltration, reverse osmosis, treatment, recycling, dyeing

4.1   Introduction

There are 57 textile units processing under large scale industries. Apart from this, cotton textiles 5156 units, hosiery and readymade garments 4799 units are operated under small scale industries in Erode District in Tamil Nadu. Water is essential to all forms of life. Erode is well-known textile centre in India particularly for textile processing. The main water resources for these industries are Bhavani and Cauvery rivers. Most of the textile industries situated in these regions are small and medium, which are unorganized in nature in view of effluent treatment facilities. Most of the colour effluents are discharged into the river particularly into river Cauvery, without any proper treatment. So it is necessary to find low-cost and affordable treatment for the colours textile wastewater.[1] In Tirupur, presently there are 712 dyeing and bleaching industries that generate 87,000 m³/day of wastewater. Out of this, a total of 281 industries are attached with common effluent treatment plants (CETP) and others are having their individual effluent treatment plants. Presently adopted technology is able to remove the colour and other organic impurities to the stipulated standards but failed to arrest the inorganic contaminants.

Continuance of effluent discharges has caused gross damages to the nearby aquatic systems receiving body like Orathupalayam dam located at the downstream of river Noyyal and as such the water quality has become unfit for irrigation. The reservoir water TDS, chloride and sodium were reported as high as 5054, 2869 and 1620 mg/l, respectively.[2] Also the concentration of dissolved solids in the ground and river water is reported in the range of 5000–7000 mg/l, i.e., almost ten times higher than the desirable drinking water standard.[3]

India is the second largest export of cotton yarn. There are about 10,000 garment manufacturers and 2200 bleaching and dyeing industries in India. Majority are concentrated at Erode and Tirupur district of Tamil Nadu, Surat in Gujarat and Ludhiana in Punjab.[4]

The environmental impact of the textile industry is associated with its high water consumption as well as by the colour, variety and amount of chemicals which are released in the wastewater. Conventional treatment methods for textile wastewater are mainly physio-chemical or biological treatments. The quality of the treatment wastewater can be improved if advanced processes are combined with them.[5]

Textile wastewater is known to exhibit strong colour, a large amount of suspended solids, highly fluctuating pH, high temperature and high COD concentration. The main pollutants in textile wastewater originate from dyeing and finishing steps which involve dyeing of the manmade or natural fibres to the desired permanent colour and processing of these into final commercial products.[6]

Textile wastewater may include many types of dyes, detergents, insecticides, pesticides, grease and oils, sulphide compounds, solvents, heavy metals, inorganic salts and fibres—amount varies depending on the processing regime. Colour removal of effluent from the textile dyeing and finishing operation is becoming important because of aesthetic as well as environmental concerns.[7]

A major environmental problem facing textile dyeing industry is that the industry produces large volume of high strength aqueous waste continuously. The industry losses valuable chemicals, consumes and water in dyeing process.[8]

4.2   Literature review

Treatment of wastewater effluent from the secondary wastewater treatment plant of a dyeing and finishing mill is investigated for possible reuse. The treatment system employed in this study consists of an electrochemical method, chemical coagulation and ion exchange. The electrochemical method and chemical coagulation are intended primarily to remove colour, turbidity (NTU) and COD concentration of the wastewater effluent while ion exchange is employed to further lower the COD concentration and reduce Fe ion concentration, conductivity and total hardness of the wastewater. To enhance the efficiency of electrochemical method, addition of a small amount of hydrogen peroxide is found to be highly beneficial. Experimental results throughout this study have indicated that the combined chemical treatment methods are very effective and are capable of elevating the water quality of the treated waste effluent to the reuse standard of the textile industry.[9]

A spiral reverse osmosis (SRO) plant to treat 12 Mld of mine water and spent cooling water. The result is a fine with no effluent problems, a new source of water for the power station and a treatment plant which produces significantly better cooling tower water and zero liquid discharge. Process details and 6 months of operational data are presented which demonstrate that good pre-treatment and cleaning system design allow SRO to produce consistent high-quality water from this difficult and varying feed. Effective use has been made of all liquid streams providing both mine and cooling water with "zero liquid discharge" status. The effectiveness of fouling resistant SRO membranes has been demonstrated on a large-scale industrial application. This now allows organic and inorganic effluents to be considered for process water reuse duties.[10, 11] The investigation has been carried out by treating wastewaters in pilot plant, reproducing on a smaller scale a separation system based on ultrafiltration and reverse osmosis. Significant indications for the exploitation of this approach on the fouling industrial scale were gained during the work. The effluent from dyeing and finishing plants, after activated sludge oxidation, was treated at 800 l/h by means of sand filtration, followed by a separation in an ultrafiltration membrane module. The last separation step, reverse osmosis at 8 bar pressure, produced a permeate (60% of the inlet flow) that relying on the analytical screening performed, was of much better quality with respect to process water presently in use. Reverse Osmosis treatment at 8 bar pressure produced in RO stage – I, 60% of permeate water of the inlet flow and reject of 40%. In RO stage II, permeate water in 35% and reject in 5% from the 40% reject water of inlet flow water.

4.3   Methodology

4.3.1   About SIPCOT, Perundurai common effluent treatment plant (PCETP)

The 14 textile units together formed PCETP. Each of the units has different shares in the treatment plant and consequently they are allowed different maximum flows that they can discharge to the treatment plant. The treatment plant only handles industrial effluent from those 14 textile industries. PCETP can operate 3600 m^3/day wash water and 450 m^3/day dye bath.

4.3.2   Wash water treatment plant

The wash water treatment plant was opened in July 2002 which reduces COD and BOD by 40–60%. They regularly measure pH, TSS, BOD, COD and TDS. The plant has no seasonal variation as the textile industry produces the same quantity throughout the year. However, the hourly inflow varies widely in both quality and quantity. The receiving tank and the bar screens are designed for the peak flow, but the units downstream. The equalisation tank are designed for an average flow and an average quality. The energy consumption is approximately 0.9 kWh/m^3 for the treated water and the cost is Rs.12–20/m^3 treated water.

4.3.3   Pre-treatment I

The flowchart for wash water treatment plant in PCETP is shown in the Figure 4.1. A number is connected to every unit. First of all wastewater comes through the bar screen (1) into the receiving sump (2), afterwards the water is pumped 12 m to the equalisation tank (3). The water is spread out through three floating aerators in the tank. The water is then again pumped 12 m with a centrifuge pump up to the flash mixing tank (4). There the water is mixed with lime, iron sulphate and polyelectrolyte. After that the water goes into the clariflocculator (5), where the particles coagulate and sink to the bottom as sludge. The outlet water from the clariflocculator goes to the clarified effluent sump (6). The water is pumped by means of the automatic valves gravity filter feed pumps the automatic valves gravity filter (AVGF) (7). The water used for backwashing in the AVGF goes back to the receiving sump (2). After the AVGF, hydrochloric acid is added and mixed into the water with a static mixer (8) to reduce the pH to 7.5–8.5. The water then goes to the stabilisation sump (9). Afterwards the water is distributed into two parallel, identical systems. The water is pumped into an automatic carbon filter (ACF) (10). The backwashing water from the ACF goes back to the receiving sump (2). After the ACF the clean water goes through a magnetic flow meter (11), which registers TDS and pH. Finally the water is pumped with a booster pump out to the field for irrigation.

The sludge from the bottom of the clariflocculator goes to the sludge sump (12) and then further to the sludge thickener (13). After the sludge thickener the sludge can go to two different ways. Most of the sludge goes to the centrifuge (14) but before the centrifuge, more polyelectrolyte is added. The rest of the sludge goes to the drying beds (15). As a final point, the sludge is packed in sacks and stored under a roof (16) until further notice.

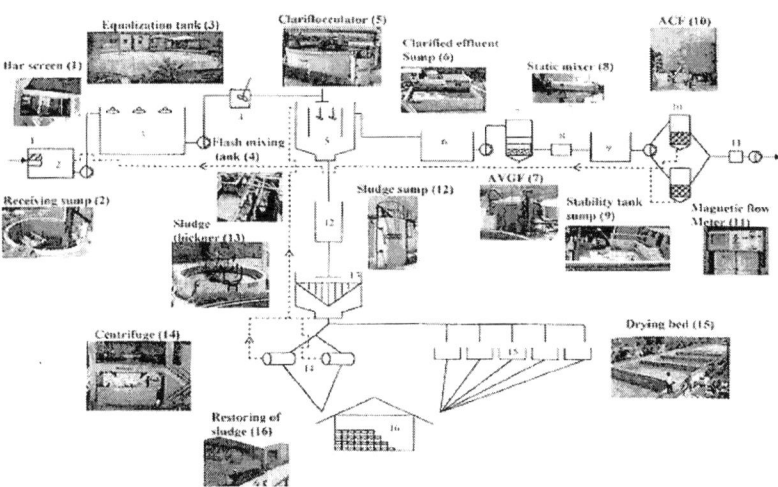

Figure 4.1. Wash water treatment plant (Pre-treatment I)

4.3.4   Pre-treatment II (Biological and Tertiary treatment)

4.3.4.1   Aeration tank/biological reactor

In the biological reactor the oxdisable organic matter is converted into carbon dioxide, water and new cells by the aerobic bacteria. The flowchart for pre-treatment II (biological and tertiary treatment), after ACF the water is pumped in to aeration tank. The tank size is 40 × 10 × 4.5 m, the next step is urea and diammonium phosphate (DAP) dosing depending upon the BOD load, use 50 kgs urea and 10 kgs DAP for 1800 m³/day.

4.3.4.2   Secondary clarifier

The mixed liquor from the biological reactor is sent to secondary clarifier for settling the biomass and a portion is pumped to biological rector for re-activation. The dimension of the tank is (20 m diameter × 3 m height). The flash mixing tank dimension is (3 × 3 × 1.5 m), from dosing tank to flash mixing tank dosing lime—1.5%, $FeSO_4$—1.25% and poly electrolyte—0.00625%. For the purpose of these materials are—lime to improve the water quality, $FeSO_4$ for colour removal and poly electrolyte for heavy metals sedimentation.

To improve the quality of secondary treated effluent,

- By reducing the suspended solids.
- By reducing the colloidal particles.
- By destroying the microorganism.
- By reducing the alkalinity.
- By reducing the colour and metal ions.

4.3.4.3   Flash mixing tank

- Dosing
- Lime – 1.5% (for improved the water quality)
- $FeSO_4$ – 1.25% (for colour removal)
- Poly electrolyte – 0.00625% (for heavy metals sedimentation).

4.3.4.4   Tertiary clarifier

Secondary clarifier overflow water is feed to tertiary clarifier. The dimension of the tank is (16 m diameter × 3 m height). Add sodium hypochlorite 3–4%. The purpose of the tertiary clarifier were as follows:

- Kill bacteria
- Reduce TSS and turbidity
- Improve water quality and clarity
- Sedimentation.

4.3.4.5   Sludge thickener

The purpose of the thickener is to increase the solids content of the sludge by removing a portion of the liquid fraction. The thickener has a slow speed mixer. The mixer has the function of making air channels in the sludge, which makes it easy for the water to escape. Another function of the mixer is to scratch the sludge into the middle of the tank where the sludge is taken out. In PCETP, the sludge thickener has a diameter of 6.0 m and a depth of 2.0 m. The dry solids (DS) after the thickener is 3–6%.

4.3.4.6   Dual media filter (DMF)

After the tertiary clarifier water is pumped in to dual media filter (DMF), water feed in DMF and passed in five stages of gravels and sand that is,

- Coarse gravels
- Fine gravels
- Coarse sand
- Fine sand
- Activated carbon (for colour removal).

The purpose of the DMF is oil and grease, silica were arrested and reduced TSS. After DMF the water is pumped into ultrafiltration (UF).

4.3.4.7   Ultrafiltration

It will remove the colloidal particles, silica, suspended solids, biological matters (bacteria, algae, fungi) and also to reduce the total organic compounds. Ultrafiltration feed water tank dimensions is (10.5 × 4.5 × 2.0 m), permeate tank dimension is (7 × 4 × 2 m). UF used in polysulfone membrane, hollow fibres and size is 0.1 μ. The purpose of the UF—colloidal particles, bacteria, suspended solids, molecular weight were reduced.

4.3.4.8   UF membrane

- Polysulfone membrane
- Hollow fibre
- Pore size 0.1 μ
- Membrane diameter 10"
- Total membrane 8 nos (modules)
- Membrane length 72".

4.3.4.9   Ultrafiltration removed

- Bacteria
- Coliform
- Virus
- Gialdia
- Cryptosporidium
- Particulate maters (metals/non-living organisms).

4.3.1   Reverse osmosis

Reverse osmosis (RO) permeate is reused for processes. The rejects about 10–12% of the inlet volume is then subjected to reverse osmosis and sent to evaporations. The final rejects from reverse osmosis system is directed to multi effect evaporator system where condensed waters are recovered.

4.3.1.1   RO membrane

- Polyamide membrane
- Spiral wound
- Pore size 5 μ
- Membrane diameter 8".

4.3.1.2   Reverse osmosis (10 lacks litre capacity)

RO – 1, 6 vessels and 4 vessels (60 membranes)
Stage – I, Feed 6 membranes (diameter 8")
Stage – II, Feed 4 membranes (diameter 8")
That is (6 vessels × 6 membrane: 4 vessels × 6 membrane)
RO – II, 2 vessels and 1 vessel (18 membranes)
Stage – I, Feed 2 membranes (diameter 8")
Stage – II, Reject 1 membrane (diameter 8")
That is (2 vessels × 6 membrane: 1 vessels × 6 membrane).
RO – I: Feed 50 m³/h
    Permeate 37.5 m³/h (75% Recovery)
    Reject 12.5 m³/h (25% Reject)
RO – II: Feed 12.5 m³/h
    Permeate 7.5 m³/h (60% Recovery)
Reject 5 m³/h (40% Reject)
RO – I and II: Feed 50 m³/h (Overall)
    Permeate 45 m³/h (90% Recovery)
    Reject 5 m³/h (10% Reject).

4.3.1.3   Reverse osmosis (26 lacks litre capacity)

RO – 1, 13 vessels and 13 vessels (156 membranes)
Stage – I, Feed 13 membranes (diameter 8")
Stage – I, Feed 13 membranes (diameter 8")
That is (13 vessels × 6 membrane: 13 vessels × 6 membrane)
High pressure pump 3 numbers.
RO – II, 4 vessels and 4 vessels (48 membranes)
Stage – I, Feed 4 membranes (diameter 8")

Stage – II, Reject 4 membrane (diameter 8")
    That is (4 vessels × 6 membrane: 4 vessels × 6 membrane).
    High pressure pumps 2 numbers.
RO – I: Feed 130 m³/h
    Permeate 97.5 m³/h (75% Recovery)
    Reject 32.5 m³/h (25% Reject)
RO – II: Feed 32.5 m³/h
    Permeate 19.5 m³/h (60% Recovery)
    Reject 13 m³/h (40% Reject)
RO I and II: Feed 130 m³/h (Overall)
    Permeate 117 m³/h (90% Recovery)
    Reject 13 m³/h (10% Reject).

4.3.1.4   RO dosing

- Dosing HCl for pH correction
- Dosing sodium meta bisulphite (SMBS) for dechlorination
- Dosing phosphoric acid base for antiscalent.

4.3.2   Dye bath treatment plant

The dye bath treatment uses an evaporator for cleaning the water. Before the evaporator the water is pre-treated in the form of sedimentation and fine screening. The evaporation unit is a high technology system that vaporises the water in five different evaporation tanks, three falling and two forced circulation (vacuum) tanks. They reduce the power input by using two heat exchangers and by doing so recover heat from the outgoing water to the incoming water. The outcomes from the evaporation tanks are two different waters, distilled water that goes back to the industries and the second water that goes to solar dryer ponds. The water in the solar dryer ponds evaporates to the atmosphere in 10 days. The rest consists to 95% of sodium chloride (NaCI). The salt is collected from the bottom of the ponds and stored in sacks under roof. They produce 3.6 tons of salt every day and the space for storage is limited so this soon becomes a big and critical issue. Purify the salt where it can be reused in the textile industries.

4.3.3   Dye bath and reverse osmosis reject treatment

4.3.3.1   Collection tank

Dye bath and RO reject is collected in collection tank and suspended solids settled down in the bottom.

4.3.3.2   Sand filter

Gravity settled dye bath and RO reject is passes through the pressure sand filter to remove suspended particles.

4.3.3.3   Evaporator

- The effluent pre-heated through five numbers pre-heaters.
- The pre-heated effluent enters into the top of the 1st calandria.
- The steam is passes through the shell side of the 1st calandria for heating the effluent.
- The vapour comes out from the 1st calandria is separated through the vapour liquid separator (VLS) and the vapour is used as a heating medium for the 2nd calandria.
- The same way the remaining calandria 3–5 are working.
- The condensate collected from the system is used for boiler and processing.
- The concentrate comes out from the system is sent to forced circulation evaporator for further concentration.
- The whole system is working under the vacuum condition.
- The feed specific gravity is 1.02 and the product specific gravity is 1.15.

4.3.3.4   Three effect forced circulation evaporator and salt recover plant

The concentrate comes out from the falling film evaporator is further concentrated from specific gravity 1.15–1.3.

The concentrate is sent to

(i)   Solar evaporation pans for salt recovery during summer days.
(ii)   Thickener for settling the salt and dewatered through the pusher centrifuge during winter.

Figures (4.2–4.15) shows different stages of treatment plant. Biological, secondary, tertiary and RO treatment.

Figure 4.2. Biological reactor

Figure 4.3. Secondary clarifier

Figure 4.4. Flash mixing tank

Figure 4.5. Dosing tank for flash mixing

Figure 4.6. Tertiary clarifier

Figure 4.7. Tertiary clarifier, sludge sump and thickener

Figure 4.8. Dual media filter

Figure 4.9. Ultrafiltration membrane

Figure 4.10. RO plant stage I and II

Figure 4.11. RO plant dosing tank

Figure 4.12. RO permeate water tank

Figure 4.13. Effluent plant overview

Figure 4.14. Multiple effect evaporator

Surface Condenser

Figure 4.15. Surface condenser

4.4   Result and discussion

Knitting fabric dyeing industry is located at SIPCOT, Perundurai, Erode, eight numbers of soft flow reactors dyeing machine (batch process), five numbers of winch dyeing are used for knitted fabric dyeing with different capacities of machines, including scouring, bleaching, mercerisation and dyeing. The process is described in Figure 4.16.

4.4.1   Sequence process for knitted fabric dyeing

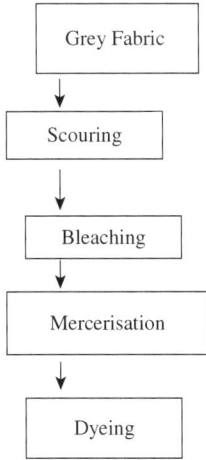

Figure 4.16. Flow chart of knitted fabric dyeing process

Scouring: To remove the natural as well as added impurities from the fabrics, i.e., oil, grease, natural impurities, natural colouring matters and mineral matters.

Bleaching: To remove the natural colouring matters and make the fabric more white.

Mercerisation: Cotton fabric treated by concentrated of caustic soda solution with high tension. This process improves the lustre, tensile strength and dye affinity.

Dyeing: Dyeing is the process of application of colours uniforming throughout the material on both sides of the fabrics with required fastness properties.

Effluents are segregated into dye bath wastewater and wash water treatment is effected accordingly. The effluent samples were collected as per APHA (American Public Health Association), AWWA (American Water Work Association) and WFE (World Federation of Exchange) standards. For this purpose seven locations have been identified which are as follows:

1. Wash water untreated effluent
2. Ultrafiltration feed parameters
3. Ultrafiltration permeate parameters
4. Ultrafiltration rejects parameters
5. Reverse osmosis feed parameters
6. Reverse osmosis permeate parameters
7. Reverse osmosis reject parameters.

Table 4.1 and Figure 4.17 shows the characteristics of wash water untreated effluent in the frequency of fifteen days.

Table 4.1. Wash water untreated effluent.

Day	pH	TDS ppm	TSS ppm	COD ppm	BOD ppm	Cl⁻ ppm	Total alkalinity ppm	Total hardness ppm
1	8.90	2200	850	705	320	830	1110	75
2	8.60	2140	840	680	316	815	1140	82
3	8.58	1830	550	670	310	695	1200	98
4	8.80	1540	300	700	340	460	1100	80
5	8.88	1610	320	680	310	700	1250	66
6	8.84	1560	300	650	330	582	1370	52
7	8.09	1910	620	690	340	860	1270	54
8	8.64	1940	750	645	310	800	1050	72
9	8.70	1680	520	700	324	680	1000	84
10	8.62	1720	390	710	320	696	1170	68
11	8.70	1590	360	690	330	510	1200	90

12	8.18	1890	530	650	340	610	1260	58
13	8.60	1690	555	680	320	640	1240	70
14	8.72	1760	410	600	310	620	1300	58
15	8.66	1740	520	600	330	660	1040	82

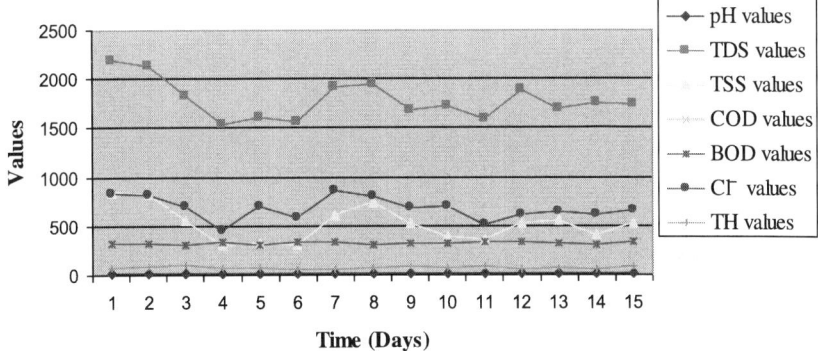

Figure 4.17. Wash water untreated effluent combined values

Table 4.2 and Figure 4.18 shows the ultrafiltration feed parameters.

Table 4.2. Ultrafiltration feed parameters.

Day	pH	TDS ppm	Cl⁻ ppm	Total hardness ppm	Total alkalinity ppm	Turbidity ppm	Free Cl_2 ppm
1	7.38	2380	1200	230	140	0.2	0.180
2	6.40	2330	1210	190	135	0.3	0.220
3	6.80	2300	1220	170	145	0.4	0.200
4	6.92	2260	1260	166	150	0.4	0.140
5	6.71	2240	1280	130	135	0.3	0.220
6	6.72	2250	1210	150	165	1.5	0.120
7	7.00	2290	1180	140	145	1.3	0.102
8	7.05	2100	1190	130	155	0.2	0.220
9	7.30	2180	1220	150	160	0.1	0.320
10	7.35	2010	1190	140	145	0.2	0.310
11	6.93	2160	1230	130	165	0.6	0.420
12	7.00	2030	1240	120	145	0.5	0.310
13	6.90	2150	1260	160	150	0.6	0.320
14	7.10	2170	1180	130	155	0.7	0.380
15	6.95	2290	1260	140	165	0.6	0.340

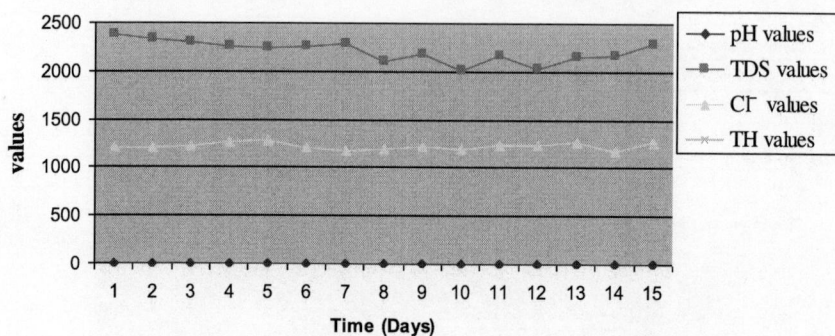

Figure 4.18. Ultrafiltration feed parameters combined values

Table 4.3 and Figure 4.19, show the ultrafiltration permeate parameters, which is the comparison between feed and permeate parameters of ultrafiltration given below.

- TDS reduced by – 9.70%
- Cl⁻ reduced by – 21.90%
- Total hardness reduced by – 56.52%.

Table 4.3. Ultrafiltration permeate parameters.

Day	pH	TDS ppm	Cl⁻ ppm	Total hardness ppm
1	6.85	2220	1140	125
2	7.15	2300	1100	150
3	7.60	2200	1090	130
4	7.20	2210	1000	150
5	6.10	2260	1100	135
6	7.25	2220	1100	135
7	7.10	2190	1110	140
8	7.20	2160	1120	128
9	7.40	2150	1100	130
10	7.20	2400	1120	140
11	7.30	2300	1110	110
12	7.36	2360	1060	100
13	7.30	2420	1090	120
14	7.20	2460	1120	130
15	7.26	2390	1140	125

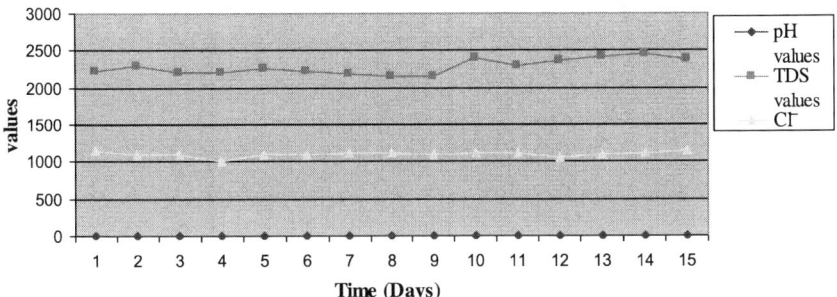

Figure 4.19. Ultrafiltration permeate parameters combined values

Table 4.4 and Figure 4.20, shows the ultrafiltration reject parameters, which is the comparison between feed and reject parameters of ultrafiltration reduction percentage given below.

- pH reduced by – 14.67%
- TDS reduced by – 17.96%.

Table 4.4. Ultrafiltration rejects parameters.

Day	pH	TDS ppm
1	7.40	2240
2	7.32	2280
3	6.80	2160
4	7.30	2150
5	7.50	2200
6	6.88	2210
7	7.35	2200
8	7.10	2100
9	7.20	2140
10	7.03	2300
11	7.40	2340
12	7.35	2390
13	7.40	2410
14	7.20	2450
15	7.30	2390

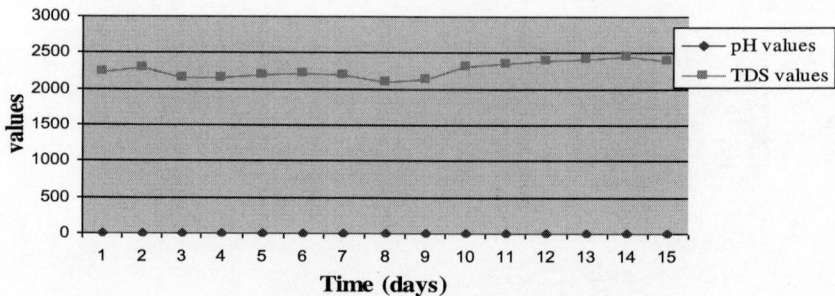

Figure 4.20. Ultrafiltration reject parameters combined values

Table 4.5 and Figure 4.21, shows the reverse osmosis feed parameters of pH, TDS, Cl⁻, total hardness, total alkalinity, SO_4, SO_3, free Cl_2 Si and Fe in the frequency of fifteen days.

Table 4.5. Reverse osmosis feed parameters.

Day	pH	TDS ppm	COD ppm	Cl⁻ ppm	Total hardness ppm	Total alkalinity ppm	SO_4 ppm	SO_3 ppm	Free Cl_2 ppm	Si ppm	Fe ppm
1	7.38	2380	60	820	180	120	230	2.4	0.105	1.56	0.07
2	6.10	2300	64	1010	170	120	240	3.2	0.070	1.77	0.05
3	7.26	2250	56	1100	160	130	248	6.2	0.109	0.52	0.26
4	6.82	2260	54	1020	152	120	220	3.1	0.054	2.62	0.09
5	6.47	2210	62	1100	132	100	230	2.6	0.060	2.20	0.11
6	6.20	2240	40	1070	142	140	310	1.6	0.084	1.94	0.10
7	6.77	2220	32	1130	134	100	280	2.2	0.084	2.20	0.37
8	6.63	2100	20	1080	142	130	230	1.3	0.037	1.68	0.13
9	6.95	2120	32	1130	138	140	230	1.8	0.065	1.74	0.07
10	6.75	2270	34	1100	136	110	260	1.6	0.080	1.60	0.13
11	6.50	2240	36	1000	132	120	240	1.9	0.071	1.69	0.18
12	6.43	2210	40	1030	130	90	300	2.4	0.080	2.20	0.13
13	7.01	2200	32	1060	128	120	240	5.8	0.072	10.70	0.10
14	7.10	2140	44	1120	120	130	260	2.2	0.056	10.55	0.05
15	6.89	2120	42	1110	114	120	290	2.7	0.062	8.25	0.06

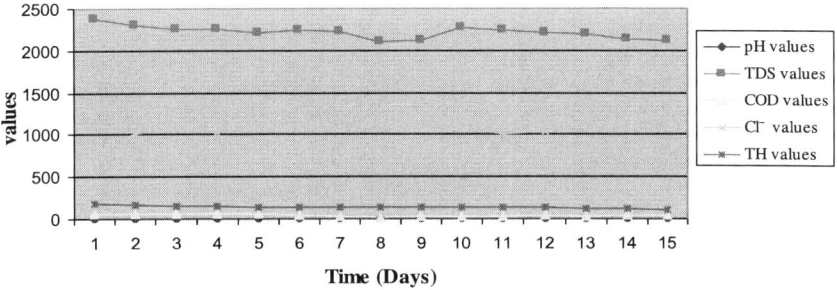

Figure 4.21. Reverse osmosis feed parameters combine values

Table 4.6 and Figure 4.22, shows the reverse osmosis permeate parameters, from the results a comparison was made between reverse osmosis feed and permeate, the reduction percentage can be given below.

- pH reduced by – 18.70%
- TDS reduced by – 96.55%
- Cl^- reduced by – 95.40%
- Total hardness reduced by – 99.17%

Table 4.6. Reverse osmosis permeate parameters.

Day	pH	TDS ppm	Cl^- ppm	TH ppm
1	6.7	88	72	5.0
2	6.6	82	60	6.0
3	6.8	90	74	6.0
4	6.9	92	80	6.0
5	6.3	70	52	2.0
6	6.4	80	62	6.0
7	6.5	90	68	6.0
8	6.3	86	61	5.0
9	6.5	82	58	2.0
10	6.2	94	70	6.0
11	6.0	98	52	1.5
12	6.2	90	55	2.0
13	6.1	100	61	1.5
14	6.2	140	82	2.0
15	6.2	130	80	2.5

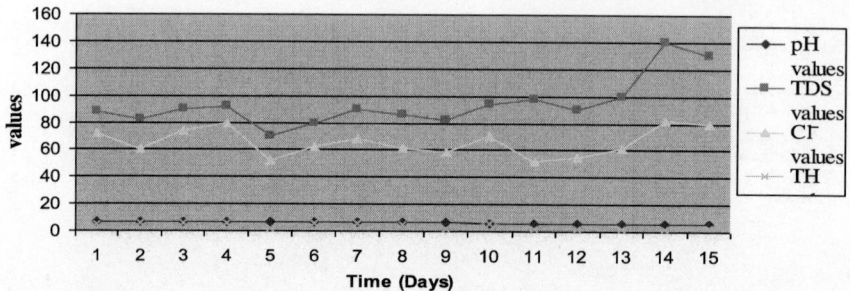

Figure 4.22. Reverse osmosis permeate parameters combined values

Table 4.7 and Figure 4.23, shows the reverse osmosis rejects parameters of pH, TDS, Cl⁻ and total hardness. The comparison between reverse osmosis feed and reject reduction percentage was given below.

- pH reduced by – 15.28%
- TDS reduced by – 90.00%
- Cl⁻ reduced by – 92.81%
- Total hardness reduced by – 91.23%.

Table 4.7. Reverse osmosis rejects parameters.

Day	pH	TDS ppm	Cl⁻ ppm	TH ppm
1	6.40	15,300	8200	1300
2	6.50	16,800	9300	1240
3	6.30	14,400	8400	1350
4	7.01	18,250	9200	1370
5	6.65	17,400	8150	1140
6	6.00	1800	7800	1200
7	7.00	21,000	8200	1250
8	6.90	18,300	9300	1290
9	7.20	20,240	8300	1250
10	6.90	17,300	8600	1210
11	6.20	16,400	8200	1130
12	6.00	12,200	5637	690
13	6.50	17,000	10,280	1210
14	7.00	16,100	11,400	1140
15	7.10	18,200	9040	1050

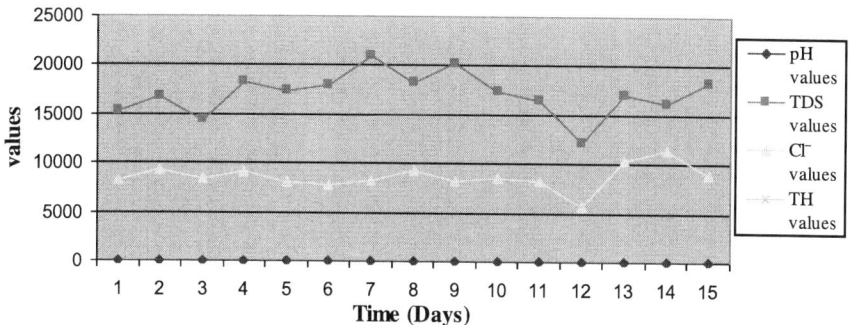

Figure 4.23. Reverse osmosis reject parameters combined values

4.5   Conclusion

The success of RO in wastewater treatment has led many industries to view this technology as a means of pollution abatement and cost savings through reuse. Applications include renovation of effluents from the various textile industries and also, treatment of landfill leachate with RO is showing promise as a means of avoiding groundwater and surface water contamination.

The values of TDS for untreated and treated and reverse osmosis, the reduction of TDS 96.55% was observed. This may be due to ability of poly membrane to capture of sodium, magnesium and calcium salts. When the effluent is passing to membrane through high pressure pump, salts of higher molecular weight are attracted by the polymembrane.

The values of COD for untreated wash water effluent, on the other hand the treated reverse osmosis, the reduction of COD 97.18%. The values of Cl^- for reverse osmosis feed and permeate, the reduced by 95.40%. The values of total hardness for reverse osmosis feed and permeate, the reduced by 99.17%.

Nomenclature

ACF	–	Activated carbon filter
AVGF	–	Auto valve less gravity filter
$FeSO_4$	–	Iron sulphate
Ca (OH)	–	Lime
DS	–	Dry solids
SS	–	Sludge sump
TNAU	–	Tamil Nadu Agriculture University
TNPCB	–	Tamil Nadu Pollution Control Board
WET	–	Water environment technology
WTC	–	Water technology centre

4.6   References

1. Malathy, R., (2007), Treatment of textile effluent using fly ash adsorbent – A case study for Tirupur region, Nature Environmental and Pollution Technology, 649–654.
2. Central Pollution Control Board, (2005), Environmental Investigations into the ground and surface water quality aspects of river Noyyal and performance evaluation of effluent treatment systems of Tirupur, Tamil Nadu.
3. Indian Standard 10500, (1991), Indian standard drinking water – specification, Bureau of Indian Standards, New Delhi.
4. Azbar, N., Yonar, T., and Kestioglu, K., (2004), Comparison of various advanced oxidation process and chemical treatment method for COD and color removal from a polyester and acetate fibre dyeing effluent, Chemosphere, 55: 35–43.
5. Bespia, A., Mendoza Roca, J.A., Roig Alcover, L., Iborra Clar, A., Iborra Clar, M.I., and Alcaina Miranda, M.I., (2003), Comparison between nanofiltration and ozonation of biological treatment textile wastewater for its reuse in the industry, Desalination, 157: 81–86.
6. Sangyong Kim, Chulhwan Park, Tak Hyun Kim, Jinwon Lee, and Seung Wook Kim, (2003), Journal of Bioscience and Bio engineering, 195: 102–105.
7. Sostar Turk, S., Simonic., and Petrinic, I., (2005), Wastewater treatment after reative priniting, Dyes and Pigments, 64: 147–152.
8. Met Baban, Ayfer Yediler, Nilgunkiran Cihz, and Antonious Kettrup, (2004), Biodegradability oriented treatability studies on high strength segregated wastewater of textile dyeing plant, Chemosphere, 57: 731–738.
9. Sheng, H.L. and Ming, L.C., (1997), Treatment of textile wastewater by chemical methods for reuse, Water Research, 31(4): 868–876.
10. Felix Buhrmann, Mike van der Waldt, Dirk Hanekom, and Fiona Finlayson, (2004), Treatment of industrial wastewater for reuse, Desalination, 124(1999): 263–269.
11. Ciardelli, G., Corsi, L., and Marcucci, M., (2000), Membrane separation for wastewater reuse in the textile industry, Resources, Conservation and Recycling, 31: 189–197.

4.7   Acknowledgement

The authors would like to thank the Perundurai Common Effluent Treatment Plant, SIPCOT, Perundurari, Erode, Tamilnadu.

5

Evolution of Antipollution Face Mask using three layer composite fabrics

Dr. M.R. Srikrishnan, Mr. J. Niresh and Dr. N. Archana,

Assistant Professor (Senior Grade), Department of Fashion Technology, PSG College of Technology, Coimbatore – 641004

Assistant Professor (Senior Grade), Department of Automobile Engineering, PSG College of Technology, Coimbatore – 641004

Assistant Professor (Senior Grade), Department of Electrical and Electronics Engineering, PSG College of Technology, Coimbatore – 641004

Abstract In recent times, India is facing an unprecedented growth of population, vehicles and small scale industries which is causing serious ecological imbalance and environmental degradation. This rapid population growth along with the high rate of urbanisation as well as industrialisation and an increase in motorised transport has resulted in an increase in the levels of various air pollutants like oxides of sulphur, oxides of nitrogen, suspended particulate matter, carbon monoxide, lead, ozone, benzene, and hydrocarbons.

Key words: bacteria, particulate matters, pollution, mask

5.1   Introduction

Delhi, in terms of air pollution, was ranked fourth among the 41 most polluted cities in the world with an annual average level of suspended particulate matter increased to 450 μg/m³, which is nearly three times the National Ambient Air Quality Standard of 140 for residential areas. As a result of increase in pollution it has led to suspended particulate matter (PM 10–PM 2.5), which causes serious respiratory and other health problems and may sometime be fatal. In order to protect the people from health hazards caused by these pollutants, it has been found necessary to develop an anti-pollution face mask which can protect the humans against these respiratory and health issues.

The commercially available face masks are capable to protect the people against these issues but still it prevails from certain drawbacks and limitations against filtration efficiency especially bacterial filtration efficiency and particulate matters. The draw backs associated with commercial face masks are price, fit and their efficiency in filtrations.

Our major objective is to develop an anti-pollution face mask which has both bacterial and particulate matter filtration at lower cost. A new attempt has been made by developing a triple layer fabric for efficient filtration purpose. The combination of fabrics that were used is spun bonded cotton and polypropylene with granular activated carbon filter fabric. The bacterial filtration efficiency of the developed product is acceptable by the standard norms by 92.8% in comparison with the surgical mask. And many of the commercially available product does not focus on the bacterial filtration efficiency, hence this feature of the developed face mask gives an added advantage and feather in the market.

5.1.1   Vehicle pollution

Nearly one half of all Americans—an estimated 150 million—live in areas that don't meet federal air quality standards. Passenger vehicles and heavy-duty trucks are the main sources of this pollution, which includes ozone, particulate matter, and other smog-forming emissions.

The health risks of air pollution are extremely serious. Poor air quality increases respiratory ailments like asthma and bronchitis, heightens the risk of life-threatening conditions like cancer, and burdens our health care system with substantial medical costs. Particulate matter (PM) is single-handedly responsible for up to 30,000 premature deaths each year.

Passenger vehicles are a major pollution contributor, producing significant amounts of nitrogen oxides, carbon monoxide, and other pollution. In 2013, transportation contributed more than half of the carbon monoxide and nitrogen oxides, and almost a quarter of the hydrocarbons emitted into the air.

Research shows that motor vehicles are responsible for about 70% of south-east Queensland's air pollution. Unless we all start reducing car use and motor vehicle pollution, this level is set to increase dramatically.

Independent research conducted by the Department of Transport and Main Roads identified air quality as a major concern among people in south-east Queensland. An average car creates almost 6 tonnes of pollutants each year. That's approximately equivalent to the weight of seven small cars. Despite this, current air quality is relatively good—but that doesn't mean it will remain that way in the future. South-east Queensland's population is predicted to reach 3 million in the next 15 years. If car use continues to rise, researchers predict we will double the number of vehicle kilometres travelled over the same period.

In recent times, India is facing an unprecedented growth of population, vehicles and small scale industries which is causing serious ecological imbalance and environmental degradation. This rapid population growth along with the high rate of urbanisation as well as industrialisation and an increase in motorised transport has resulted in an increase in the levels of various air pollutants like oxides of sulphur, oxides of nitrogen, suspended particulate matter, carbon monoxide, lead, ozone, benzene, and hydrocarbons.

Delhi, in terms of air pollution, was ranked fourth among the 41 most polluted cities in the world with an annual average level of suspended particulate matter increased to 450 µg/m³, which is nearly three times the National Ambient Air Quality Standard of 140 for residential areas. As a result of increase in pollution it has led to suspended particulate matter (PM 10–PM 2.5), which causes serious respiratory and other health problems and may sometime be fatal.

5.1.1.1   Cars and air pollution

The principal air-quality pollutant emissions from petrol, diesel, and alternative-fuel engines are carbon monoxide, oxides of nitrogen, un-burnt hydrocarbons and particulate matter. It is emissions of these pollutants that are regulated by the Euro emissions standards. Modern cars, if kept in good condition, produce only quite small quantities of the air quality pollutants, but the emissions from large numbers of cars add to a significant air quality problem. Carbon monoxide, oxides of nitrogen, and un-burnt hydrocarbons are gases, and are generally invisible. Particulate matter is usually invisible although under certain operating conditions diesel engines will produce visible particles, appearing as smoke. Petrol engines will also produce visible particles if they are burning engine oil or running "rich", for example, following a cold start. Fine particles can also be produced by tyre and brake wear. Pollutant emission levels depend more on vehicle technology and the state of maintenance of the vehicle. Unlike emissions of CO_2, emission of air quality pollutants are less dependent on fuel consumption. Other factors, such as driving style, driving conditions and ambient temperature also affect them. However, as a starting point, all new passenger cars must meet minimum EU emissions standards.

5.1.2   Pollution prevalence

Air pollution has remained consistently high and rising. Suspended particulate matter content has also become high. Air pollution has led to many severe health problems. A study by Health Effect Institute, Boston, estimated at least

3000 premature deaths due to air pollution related diseases. WAO journal also reported high respiratory disorder symptoms. The main problem is with heavy traffic movement and WAO alerts that allergic problems will increase further as air pollution increases.

The most detailed study of all pollution sources by IIT report includes wide range of pollution sources including direct emissions from sources as well as indirect formation of particulate from the gases in the air. The main NOX and PM 2.5 load emissions from different sources.

5.1.2.1   Nitrogen dioxide

The annual average values of NO_2 observed during last 6 years. Despite, an increase in the number of vehicles, the NO_2 levels have shown an increase from 41.7 to 47.2 mg/m^3 which is not very significant. The annual mean of NO_2 levels in Delhi is well within the annual average of National Ambient Air Quality Standards for residential areas, which is 60 mg/m^3.

5.1.2.2   Suspended particulate matter (SPM)

The SPM levels have increased from 363 to 456 mg/m^3. But, SPM levels reduced by 11.4% in the year 2001 and increased again to 19.4%. This could be attributed to the adverse meteorological conditions.

5.1.2.3   Respirable particulate matter (RSPM) or PM10

The annual average of respirable particulate matter (RSPM) levels over the city. It is clear that RSPM levels have reduced to 21.4% in the year 2001 and remained almost constant. The increase could be due to adverse meteorological conditions.

5.1.2.4   Carbon monoxide (CO)

The annual average carbon monoxide (CO) levels, indicate constant reduction of CO levels at ITO (Indium tin oxide) intersection. The observed concentration of CO came down from 4183 mg/m^3 to 3258 mg/m^3. This could be attributed to the stringent vehicular emission norms, fuel quality up-gradation and development of better engines. Though the annual average levels of carbon monoxide continue to be above the danger mark of 2000 mg/m^3, there has been a gradual decline.

5.1.2.5   Lead

The annual average levels of lead in the city which have substantially reduced. The lead concentration in petrol was brought down from 0.56 g/l to 0.15 g/l. The lead was totally phased out from petrol. These measures resulted in the reduction of the lead levels in the ambient air (Figure 5.1).

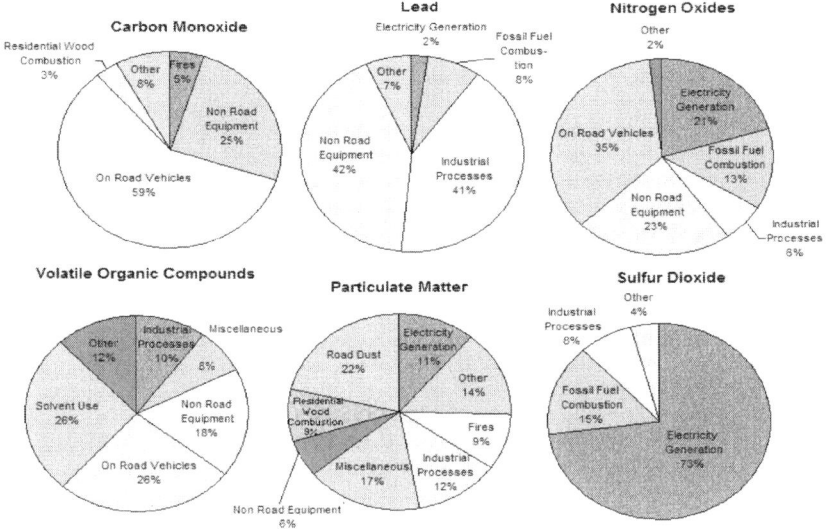

Figure 5.1. Pollutants

5.2   Health impacts

The people of Delhi still have memories of the Bhopal disaster. The Bhopal Gas was caused by an acute exposure to methyl-isocyanate (MIC) leaking from the Union Carbide pesticide plant and killed about 2000 people. This disaster is a tragic illustration of the impact of pollutants on human health.

The health effects of pollutants depend upon the concentration, exposure duration and the individual's susceptibility. In general, after an initial lag period the health effects become manifest and continue to rise, reaching a plateau thereafter. There are important connections between air pollution and diseases, and the cost that they impose on the society.

A positive, significant relationship between particulate pollution and daily non-traumatic deaths as well as deaths from certain causes (respiratory and cardiovascular problems) and for certain age groups. In general, these impacts are smaller than those estimated for other countries, where on an average a 100-μg increase in total suspended particulates (TSP) leads to a 6% increase in non-traumatic mortality. In Delhi, such an increase in TSP is associated with a 2.3% increase in deaths.

The differences in magnitudes of the effects are most likely explained by differences in distributions of age at death and cause of death, as most deaths in Delhi occur before the age of 65 and are not attributed to causes with a strong association with air pollution. Although air pollution seems to have less impact on mortality counts in Delhi, the number of life years saved per death avoided is greater in Delhi than in US cities because the age distribution of impacts in these two places varies.

In US particulates have the greatest influence on daily deaths among persons 65 and older. In Delhi, they have the greatest impact in the 15–44 age group. That means that for each death associated with air pollution, on average more life-years would be saved in Delhi than in US.

Large differences in the magnitude of effects do call into question the validity of the "concentration-response transfer" procedure. In that procedure, concentration-response relationships found for industrial countries are applied to cities in developing countries with little or no adjustment, to estimate the effects of pollution on daily mortality.

The World Health Report attributes environmental risks especially the urban air pollution, indoor air pollution, lead exposure and climate change as some of the causes for the Disability Adjusted Life Year (DALY).

With the multi-pronged efforts taken by the Delhi Government in recent years as already mentioned, the concentration of both SO_2 and CO have declined. However, the impact of introduction of CNG on the levels of NOX, SPM and RSPM needs to be studied in detail.

No specific epidemiological study on impact of air pollution and its effects on human health has been done in Delhi. More importantly, cost-effective solutions need also to be developed through advanced research and analysis and integrated into the policy framework in various sectors like transport, health and even the industrial policy. This has not happened so far (Figures 5.2, 5.3).

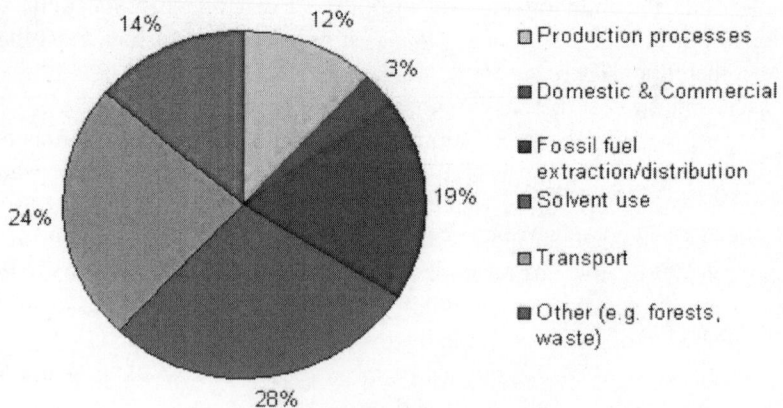

Figure 5.2. Sector wise pollution

Figure 5.3. Health impacts

5.3   Mask

Face masks that filter out airborne particles can make a big difference in the exposure to air pollution. A mask with a high quality filter that fits properly can be an effective measure against the inhalation of harmful pollutant particle as small as PM 2.5.

Not all masks are effective against breathing in small pollutant particles. Therefore mask with proper certification and ratings are important. They indicate that the mask has been tested and meets benchmark standards to filter out small airborne particles (Figure 5.4).

Always make sure that mask fits securely and there are no gaps to let outside air. A gap in a mask allows air in completely negates any benefit or protection that would get from filtering out pollutant particles.

Then to be sure to check the mask's material to ensure that it can filter out small particles. The mask must always be well ventilated.

Figure 5.4. Commercial face masks

5.3.1   Existing market product analysis

1. Vogmask
   - Eco-friendly
   - Reusable mask
   - Fits well on the face
   - Prevents breathing free of pollutants
   - Comfortable to wear even with helmets.
2. Respra
   - Soft and easy to adjust
   - Gives a comfortable and close fit
   - Usable for four months
   - Ideal for motorists, cyclists and pedalist.
3. Breathe sportive mask
   - Replaceable filters
   - Durable and washable
   - Easy breathing
   - Don't get fogged-up.
4. V-kare
   - Helps to avoid pollution
   - Comfortable to wear
   - Prevents all kinds of infection.

5.   On-mask
  • Adjustable elastic loops
  • Gives excellent seal and nose fit
  • Available in large, medium and small sizes
  • Light weight and comfortable to wear
  • Washable and reusable.

Drawback
  • Effective and quality masks are costlier
  • Difficulty in breathing (not enough ventilation)
  • It is not smoke resistant
  • Its allergic to some people
  • It does not fit well and hence leakages occur
  • They do not kill any microbes
  • The contaminant sits on top till it gets disposed
  • Ozone, vapours and nitric oxides are not blocked
  • It is not available for all sizes.

5.3.2   Materials and methods

In this chapter, the materials to be used in the development and its properties which qualify it to use is discussed. And also the processing and test methods are discussed in a detailed way (Figure 5.5).

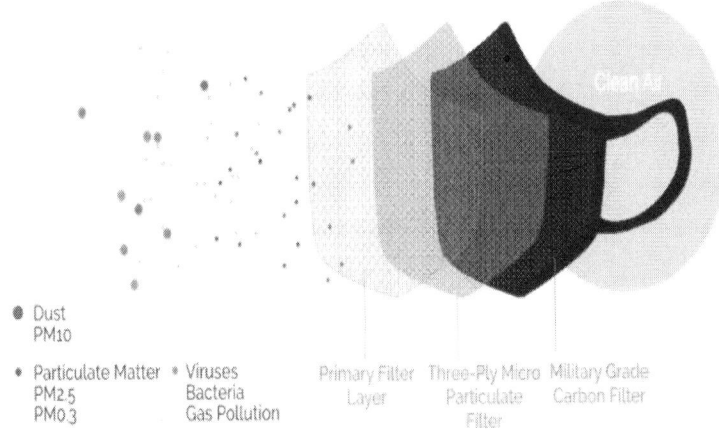

Figure 5.5. Overview of three layer mask

5.3.2.1   Methodology

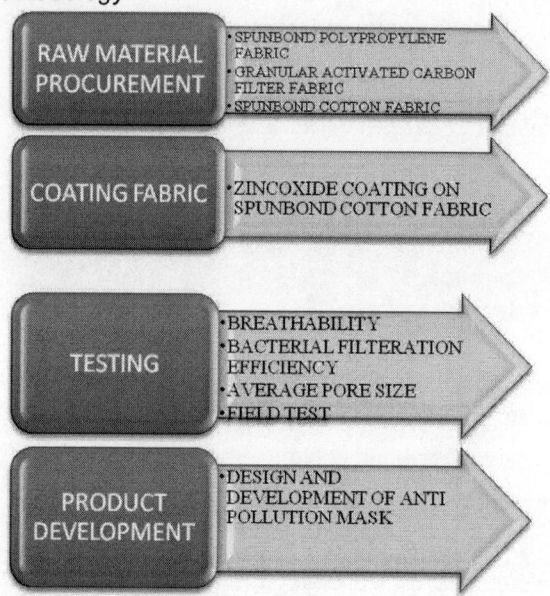

5.3.2.2   Materials

For the production of anti-pollution mask (Figure 5.6).

- Outer layer—spun bond polypropylene fabric
- Filtration layer—granular activated carbon filter fabric
- Inner layer—spun bond cotton fabric coated with zinc oxide.

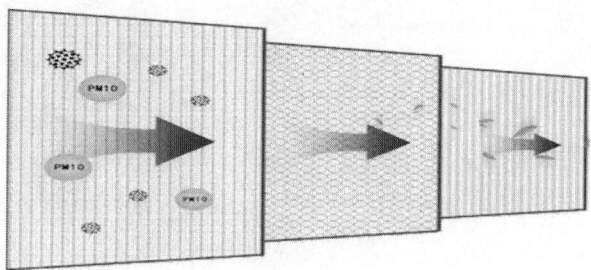

Figure 5.6. Direction of filtration

5.3.2.3   Non-woven polypropylene fabric

Technically polypropylene (PP) is a plastic noted for its light weight, being less dense than water; it is a polymer of propylene. It resists moisture, oils, and solvents. Since its melting point is 121°C (250°F), it is used in the manufacture of objects that are sterilised in the course of their use. Polypropylene is also used to make textiles, ropes that float, packaging material, and luggage.

Since recent years, it tends to replace the traditional existing textiles, fabric and paper for some obvious reasons. The main asset of non-woven polypropylene (PP) fabric is that it is made of spun bond polypropylene which can be recycled, naturally decomposes (untreated nonwoven fabric can decompose in the outside within a few months only) and completely incinerates without any production of poisonous pollutants (does not pollute directly or indirectly the environment). It does not contain any harmful substances itself and does not require the use of poisonous gas, wasted oil or effluents (Figure 5.7).

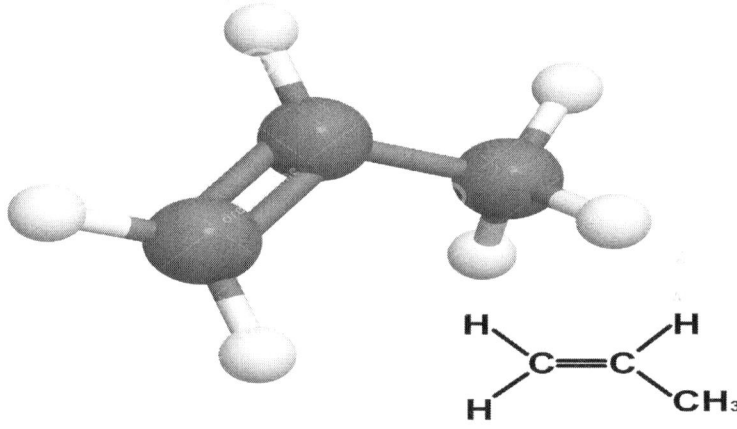

Figure 5.7. Structure of polypropylene

It is a polymer of ethylene, and is produced at high pressures and temperatures in the presence of any one of several catalysts, depending on the desired properties for the finished product, eco-friendly PP can go a long way for the concept of environmentally friendly products and can be expanded when you choose to use non-woven PP fabric.

Non-woven fabrics got environmental benefits over other traditional fabrics and papers in regards to mainly its production process and its recycling benefits due to natural degrading properties and the recycle process for its production. Cheap, non-woven is very competitive with other fabrics (often cheaper than paper or plastic bag) and is very durable with all the attribute of woven fabrics as mentioned above (softness, air permeability, dehumidifying, cushioning, resilience, good light weight, ability to repel water and to evaporate water, resistance to mould and insect., etc.).

Various add-on treatment can be added to the PP fabric such as (Figure 5.8):

- Flame retardency (for Airplanes, Hotels designated products)
- Anti-bacterial (medical use products)
- Florescent treatment (caution, promotion and advertising items)
- Compounded treatments (adding layers of others fabric for various purpose such as aluminium film, PE film, Eva film, PVC, absorbent paper, CPP film, etc.)
- Anti-mildew treatment (storage purpose)
- Lamination treatment for offset printing, etc.

Figure 5.8. Polypropylene fabrics

5.4   Filtration theory of airborne particles

Media used for the filtration of airborne particles do not work by the same principles as those used for the filtration of liquids. Filters used in respirators and medical masks must allow the user to breathe and thus cannot clog when particles adhere to their fibres. Respirator and medical mask filters are typically composed of mats of nonwoven fibrous materials, such as wool felt, fiberglass paper, or polypropylene. The material creates a tortuous path, and various mechanisms result in the adhesion of particles to the fibres without necessarily blocking the open spaces, still allowing air to flow easily across the filter.

Three mechanisms of removing particles from the airstream: inertial impaction, diffusion, and electrostatic attraction. Mechanisms for removing large particles differ from those for small particles. The model postulates that inertial impaction is effective for aerosol particles that are approximately 1 µm and larger. Such particles have enough inertia that they cannot easily flow around the respirator fibres. Instead of flowing through the filter material, the large particles deviate from the air streamlines and collide with the fibres and may stick to or be caught in them.

For much smaller particles—those that are 0.1 μm and smaller—diffusion is regarded as an effective filtration mechanism. Brownian motion—the process by which the constant motion of oxygen/nitrogen molecules causes collisions between particles—results in a "wandering" pathway. The complex path that is followed by the small particles increases the chance that they will collide with the filter fibre and remain there.

Another efficient method of capturing both large and small particles from the airstream is said to be electrostatic attraction, in which electrically charged fibres or granules are embedded in the filter to attract oppositely charged particles from the airstream. The attraction between the oppositely charged fibres and particles is strong enough to effectively remove the particles from the air. The first electrostatic filters used resins.

The filtering materials of respirators and medical masks are typically non-woven. These materials, initially using natural fibres, came into greater prominence with the introduction of synthetic thermoplastics, particularly polypropylene, about 40 years ago. Spun-bonded polypropylene is a fabric or structure in the category of non-woven textile materials. The salient advantage of non-woven technology is the ability to produce fabrics or structures at significantly lower cost than the older fabric-generating techniques of weaving or knitting of spun yarns. Additional important advantages are the versatility of the process and the products in terms of properties and uses. There has been an on-going development of and increasing sophistication in spun-bonded, and the related melt blown, technologies, which have made these materials the optimal choice in many applications.

Polypropylene is one of five major commodity plastic resins now produced in large quantities in many countries. It is readily converted into spun-bonded fabric and structures with a very wide range of properties. Some of the parameters that can be varied include fibre thickness (down to micron or submicron diameters), density of fibres per unit area or volume, density of bond points, and average orientation of fibres.

For filtration and trapping of aqueous particles (as in respirators and medical masks), polypropylene fibre surfaces require modification to render them more hydrophilic (water attracting) because polypropylene is inherently hydrophobic (water repelling). Several methods are known to impart the necessary degree of hydrophilicity to the surface. A process in which a droplet-attracting electric charge is applied to the surface has also been described, but it is not clear that such a charge could be maintained during storage of the respirator or mask, and the charge would dissipate with exposure to air with any degree of humidity. The efficiency of resin electrostatic filters is degraded when they are exposed to airborne oil mists and other materials that shield the electrostatic charge.

Manufacturers have been able to overcome this issue by incorporating synthetic plastic fibres, such as polypropylene, which are said to be capable of holding a sufficiently strong electrostatic charge (electret) to effectively resist the shielding effects of oil.

Once particles are captured by a filter, they are held tightly to the fibres through van der Waals bonding and other forces, thus making it difficult for captured particles to escape. Filters generally become more efficient with loading (i.e., the adhesion of additional particle to the filter fibres). This increase in efficiency is the result of the increased number of collection points that are created by the particles that have already adhered to the filter fibres. However, increased loading becomes a problem when enough particles have been captured to begin to block the open spaces of the woven or non-woven network. This blockage results in a build-up in pressure drop and an increase in resistance that eventually makes it difficult to breathe while wearing the respirator. Heavy loading of filters may also increase the ability to dislodge particles that have already been captured. Very little research has been conducted on the characteristics of filters in relation to loading. However, the relatively clean environment in healthcare facilities and the limited time of use of a respirator suggest that filter clogging will rarely become an issue. Loading might be of some concern for use in areas that have considerably dirtier air than healthcare facilities.

5.4.1   Granular activated carbon filter fabric

Granular activated carbon (GAC) is a hybrid mixture of a wide variety of graphite platelets that are interconnected by non-graphitic carbon bonding. The adsorptive capacity of GAC makes it ideal for removing a variety of contaminants from water, air, liquids and gases. GAC is also an environmentally responsible product that can be reactivated through thermal oxidation and used multiple times for the same application.

The US Environmental Protection Agency (EPA) and most state-based departments of health consider adsorption by GAC to be the best available technology for the removal of many organic materials in surface water. On its own or paired with an ultraviolet (UV) disinfection system, GAC can facilitate the removal of:

- Disinfection by-products (DBPs) associated with chlorine and alternative disinfectants

- Algal toxins, such as *microcystin-LR, cylindrospermopsin* and *anatoxin-A*

- Endocrine-disrupting compounds
- Pharmaceuticals and personal care products
- Taste and odour-causing compounds
- Organic materials from decaying plants and other naturally occurring matter which serve as the precursors for DBPs.

GAC has also shown to be an effective physical filtration medium in water treatment plants and has the added benefit of providing water quality protection by adsorption of taste and odour compounds or chemical contaminants.

Prior to retrofitting an existing multimedia filter with GAC, or designing a new GAC filter, several practical considerations are necessary, including hydraulic requirements, filter on-stream time, and backwash water availability. The properties of the GAC, such as adsorption performance, abrasion resistance and density must be considered as well. Additionally, the effective cost of converting the filter to GAC must be evaluated.

Backwashing requirements are another factor affecting how deep the GAC bed should be, particularly when retrofitting a multimedia filter. The filter design should allow for the optimal expansion while allowing several inches of clearance between the top of the expanded bed and the bottom of the backwash trough. Since the water velocity increases between the troughs, any GAC that is expanded to the level between the troughs will more than likely be washed out of the filter.

The mechanics of installing GAC in a filter retrofit or in a new filter are much the same as for installing sand and anthracite. A notable exception is that the filter and initial gravel/sand or underdrain/sand support must be disinfected prior to installing the GAC. This is because suitable disinfectants, such as chlorine, will rapidly react over GAC, leaving no disinfectant residual. A second variance is the GAC should be submerged for 24 h prior to the final backwash before bringing the new filter into service. This soak time allows the air entrapped within the GAC pore structure to be removed and it allows the water to thoroughly wet the GAC internal surface (Table 5.1).

Effectiveness measured by relative adsorption rate on scale of 1-10 (10 being the best)

Table 5.1. Effectiveness of gas/vapour.

Gas / vapour	Effectiveness
VOC's (50–100)	9
Nitrogen dioxide	8
Sulphur dioxide	7
Ozone	10
Carbon oxides	0

5.4.1.1   Activated carbon explained

Activated carbon is carbon that has been treated with oxygen; this causes millions of tiny pores to open up on the carbon's surface. In fact, these pores are so numerous that a single pound of activated carbon may provide 60–150 acres of surface area to trap pollutants. Once carbon has been activated, it can remove a long list of airborne chemicals, including alcohols, organic acids, aldehydes, chlorinated hydrocarbons, ethers, esters, ketones, halogens, sulphur dioxide, sulphuric acid, and phosgene, among many others. It also removes odours, whether they are from humans or animals. It removes perfumes, household cleaning chemicals, and is especially good at removing volatile organic compounds (VOCs). Sometimes the carbon is treated with oxygen and another chemical, often a potassium. This creates a chemisorbent, which is better at removing inorganic gases and gases that are highly chemically-reactive.

How activated carbon works

Activated carbon works by adsorption—the process by which a gas bonds to the surface of a solid. In this case, the solid is the activated carbon, which adsorbs as much as 60% of its weight in airborne pollutants. Air passes through the filter where airborne gases, chemicals, and odours react chemically with the surface of the carbon, effectively sticking to it. The clean air then flows out of the filter. Activated carbon filters will adsorb even a small amount of almost all vapours and they have a large capacity for removing organic molecules like solvents. They can also simultaneously adsorb many different kinds of chemicals, making these filters very efficient. Activated carbon works in any temperature or humidity and isn't toxic, so it's safe for people to handle. Thankfully, it's also quite affordable.

Activated carbon filters effectively work like a sponge. The more activated carbon in the filter, the more pollutants it can remove and the longer the filter lasts. The best filters include many pounds of activated carbon to ensure a longer life before needing replacement.

Benefits of activated carbon filters

Activated carbon filters are the most effective way to remove gases, chemicals, cigarette smoke, and other odours from the air. They are especially helpful for those with multiple chemical sensitivity or any respiratory condition that makes one sensitive to airborne chemicals. Many high-quality air purifiers contain both HEPA (High efficiency particulate air) and activated carbon filters. Air purifiers may offer several different types of activated carbon filters, treated with different chemicals and thereby targeting specific chemicals. For example, certain air filters are specifically meant to target ammonia gases. Others target VOCs. These are not the only pollutants that the purifiers remove, but they are specially designed to be very effective at removing the target pollutant (Figure 5.9).

Figure 5.9. Carbon fibre fabric

5.4.2   Spun bond cotton fabric

Cotton is a soft, absorbent and breathable natural fibre worn close to the skin. Cotton keeps the body cool in summer and warm in winter because it is a good conductor of heat. Cotton is non-allergic unlike synthetic fibres, cotton fibre is a natural product that contains no chemical. Cotton, due to its unique fibre structure breathes better and its more comfort than oil based synthetic fibres. Cotton is one of the easiest fabric to dye due to its natural whiteness and high rate of absorbency. Cotton hold up to 27 times its own weight in water and becomes stronger when wet. Cotton can't hold an electrical charge. The strength of the cotton fibre is good. The drapability of cotton fibre is awesome. We can use the cotton fibre made fabric in any kind of wear which needs more flexibity and drape. Cotton fibre has large amorphous portion and this is why air can be in and out through cotton fibre. So, the fabric made by cotton fibre is quite comfortable to use. Cotton fibre is too much regular fibre and if properly ginned this fibre can be the best soft hand feeling fibre amongst the others. Cotton fibre has high absorbency power and this is why this fibre can be dyed properly. Good colour retention. If printing applied on cotton it seems it does not spread the colour outside the design. So, printing efficiency is good on cotton fibre (Figure 5.910).

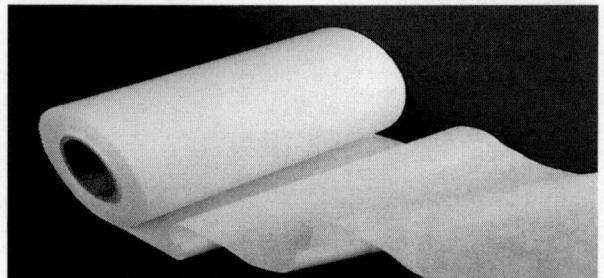

Figure 5.10. Spun bonded cotton fabric

5.4.3   Zinc oxide coating

Zinc oxide is described as a functional, strategic, promising, and versatile inorganic material with a broad range of applications. It is known as II–VI semiconductor, since Zn and O are classified into groups two and six in the periodic table, respectively. Zinc oxide holds a unique optical, chemical sensing, semiconducting, electric conductivity, and Piezo-electric properties. It is characterised by a direct wide bad gap (3.5 eV) in the near-UV spectrum, a high excitonic binding energy (50 meV) at room temperature, and a natural n-type electrical conductivity. These characteristics enable Zinc Oxide to have remarkable applications in diverse field. The wide band gap of Zinc Oxide has significant effect on its properties, such as the electrical conductivity and optical absorption. The excitonic emission can persevere higher at room temperature and the conductivity increases when ZnO doped with other metals. Though ZnO shows light covalent character, it has very strong ionic bonding in the ZnO. Its longer durability, higher selectivity, and heat resistance are preceded than organic and inorganic materials. The synthesis of nano sized ZnO has led to the investigation of its use as new antibacterial agent. In addition to its unique antibacterial and antifungal properties, ZnO-NPs possess high catalytic and high photochemical activities. ZnO possesses high optical absorption in the UVA (315–400 nm) and UVB (280–315 nm) regions which is beneficial in anti-bacterial response and used as a UV protector in cosmetics.

ZnO exhibits three crystallize structures namely, wurtzite, zinc-blende and an occasionally noticed rock-salt. The hexagonal wurtzite structure possesses lattice spacing a = 0.325 nm and c = 0.521 nm, the ratio c/a * 1.6 that is very close to the ideal value for hexagonal cell c/a = 1.633. Each tetrahedral Zn atom is surrounded by four oxygen atoms and vice versa. The structure is thermodynamically stable in an ambient environment, and usually illustrated schematically as a number of alternating planes of Zn and

O ions stacked alongside the c-axis. Zinc-blende structure is meta-stable and can be stabilised via growth techniques. These crystal structures are illustrated in, and the black and grey-shaded spheres symbolize O and Zn atoms, respectively.

Anti-bacterial activity of zinc oxide
Bacteria are generally characterised by a cell membrane, cell wall, and cytoplasm. The cell wall lies outside the cell membrane and is composed mostly of a homogeneous peptidoglycan layer (which consists of amino acids and sugars). The cell wall maintains the osmotic pressure of the cytoplasm as well the characteristic cell shape. Gram-positive bacteria have one cytoplasmic membrane with multilayer of peptidoglycan polymer, and a thicker cell wall (20–80 nm). Whereas gram-negative bacteria wall is composed of two cell membranes, an outer membrane and a plasma membrane with a thin layer of peptidoglycan with a thickness of 7–8 nm. NPs size within such ranges can readily pass through the peptidoglycan and hence are highly susceptible to damage (Figure 5.11a–c). The cytoplasm, a jelly-like fluid that fills a cell, involves all the cellular components except the nucleus. The functions of this organelle

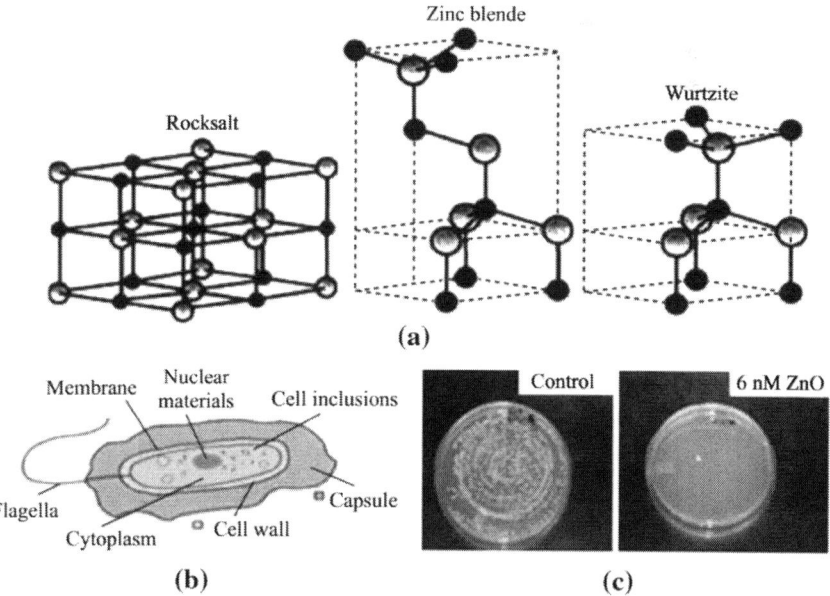

Figure 5.11. (a) ZnO crystal structures. Adapted from Ozgur et al. (b) Bacterial cell structures, reused from Earth Doctor, Inc., formerly Alken- Murray. (c) *S. aureus* plating for colony count.

include growth, metabolism, and replication. Consequently, the cytoplasm contains proteins, carbohydrates, nucleic acids, salts, ions, and water (*80%). This composition contributes in the electrical conductivity of the cellular structure. The overall charge of bacterial cell walls is negative. Figure 5.11b shows typical bacteria cell structure. Anti-bacterial activity is known according to The American Heritage Medical Dictionary 2007, as the action by which bacterial growth is destroyed or inhibited. It is also described as a function of the surface area in contact with the microorganisms. While antibacterial agents are selective concentration drugs capable to damage or inhibit bacterial growth and they are not harmful to the host. These compounds act as chemo-therapeutic agents for the treatment or prevention of bacterial infections (Saunders Comprehensive Veterinary Dictionary) (Figure 5.12a,b).

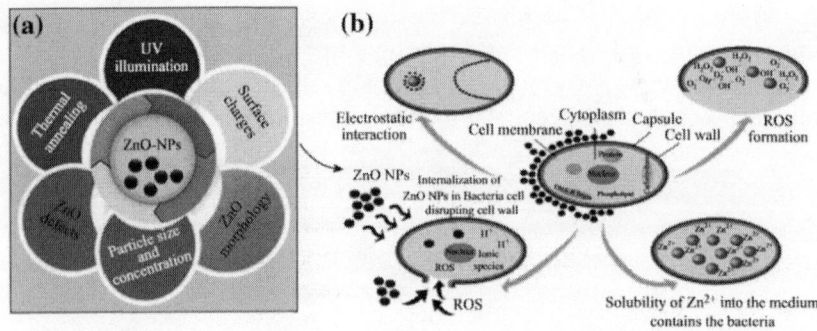

Figure 5.12. (a) Correlation between (a) the influence of essential ZnO-NPs parameters on the anti-bacterial response and (b) the different possible mechanisms of ZnO-NPs anti-bacterial activity, including: ROS formation, Zn^{2+} release, internalisation of ZnO-NPs into bacteria, and electrostatic interactions

The anti-bacterial agent is considered as bactericidal if it kills bacteria or as bacterio-static if it inhibits their growth. Different methods have been adopted for the assessment and investigation of anti-bacterial activity in vitro. These methods include disk diffusion, broth dilution, agar dilution, and the micro-titre plate-based method. Other methods are different according to the investigated parameters. For example, the conducto-metric assay measures the bacterial metabolism-induced alterations in the electrical conductivity of growth media. The most commonly used method is the broth dilution method, followed by colony count, through plating serial culture broths dilutions which contained ZnO-NPs and the targeted bacteria in appropriate agar medium and incubated. A number of researchers have examined the anti-bacterial activity of ZnO-NPs to determine bacterial growth through the culture turbidity and the viable cells percentage by the colony counts test. While others, such as Yamamoto enhanced the anti-bacterial activity of ZnO-NPs by modulating within the procedure. They considered that the anti-bacterial activity rate was much improved by decreasing the initial number of bacterial cells from 102 to 106 colony forming unit (CFU). Considered that the determination of starting number of bacterial cells is very important in the antibacterial activity evaluation. The minimum inhibitory concentration of an anti-microbial agent and minimum bacterial concentration can be measured by using the susceptibility test methods. However, there are some variations in the established laboratory methods and protocols in the assessment of the bactericidal activity. The agar diffusion method (an indirect method) is the most frequently used method and has been standardised as an official method for detecting bacterio-static activity by the American type culture collection. Other direct test methods, such as the measurement of urease inhibition of inocula, have been reported. The micro-dilution method is a modification of the broth macro-dilution test, which utilises the advances in miniaturisation to allow multiple tests to be performed on a 96-well plate. Modified procedures along with the standard methods are also used by a large body of researchers. Growth curves were typically obtained via monitoring the optical density (OD), at wavelength of 600 nm, a typical wavelength for cells. The density of bacterial isolates must be adjusted to an optimal density of 0.5 McFarland standards. The OD should serially be monitored hourly up to 12 h of incubation, and finally after 24 h of overnight incubation for the determination of the percentage of growth inhibition. The inhibition rate varies with the tested organisms and the utilised NP-oxide. We discuss below the influence of essential physiochemical and structural factors, which affect the antibacterial activity of ZnO-NPs, and consequently have potential impact upon the resultant toxicity mechanism.

5.5   Material specifications

1. **Polypropylene non-woven fabric**
   - Fabric weight: 30 GSM
   - Thickness: 0.40 mm
   - Spun bonded polypropylene is preferred for its light weight
   - It has excellent water repellency
   - It has excellent air permeability
   - It provides comfort and dryness
   - It has anti-fungal resistance.

2. Activated carbon filter fabric

- Fabric weight: 110 GSM
- Greater adsorption capability and efficiency
- Filter media for removal of acid, base and VOCs
- Low-pressure drop
- Non-wove media to arrest particulate matter
- Fill and finish both soft and flexible and stiff finish.

3. Cotton non-woven fabric

- Fabric weight: 11 GSM
- Thickness
- Soft and comfort
- Non-allergic
- Breathable
- High rate of absorbency.

5.5.1   Testing methods

5.5.1.1   Bacterial efficiency ASTM F 2101

Standard test method for evaluating the bacterial filtration efficiency (BFE) of medical face mask materials, using a biological aerosol of *Staphylococcus aureus*

- This test method is used to measure the bacterialfitration efficiency (BFE) of medical face mask materials employing a ratio of the upstream bacterial challenge to downstream residual concentration to determine filtration efficiency of medical face masks materials.

- This test method is a quantitative method that allows filtration efficiency for medical face mask materials to be determined.

- The filtration efficiency that can be determined by this method is 99.9%. This test method does not apply to all forms or conditions of biological aerosol exposure.

- Users of the test method should review modes for worker exposure and assess the appropriateness of the method for their specific application.

- This test method evaluates medical face mask materials as an item of protective clothing but does not evaluate materials for regulatory approval as respirators.

- If respiratory protection for the wearer is needed, a NIOSH certified respirator should be used.

- Relatively high bacterial filtration efficiency measurement for a particular medical face mask material does not ensure that the wearer will be protected from biological aerosols.

- Since this test method primarily evaluates the performance of the composite materials used in the construction of the face mask and not its design, fit or facial sealing properties.

The values stated in SI units or inch-pound units are to be regarded separately as standard. The values stated in each system may not be exact equivalents. Therefore, each system shall be used independently of the other. Combining values from the two systems may result in non-conformance of the standard. This test method does not address breathability of the medical face mask materials or any other properties affecting the ease of breathing through the medical face mask material.

This standard does not purport to address all of the safety concerns, if any associated with its use. It is the responsibility of the user of this standard to establish appropriate safety and health practices and determine the applicability of regulatory limitations prior to use.

Summary of test method
The medical face mask material is clamped between a six-stage cascade impactor and an aerosol chamber. The bacterial aerosol is introduced into the aerosol chamber using a nebuliser and a culture suspension of *Staphylococcus aureus* (*S. aureus*). The aerosol is drawn through the medical face mask material using a vacuum attached to the cascade impactor. The six stage cascade impactor uses six agar plates to collect aerosol droplets which penetrate the face mask material. Control samples are collected with no test specimen clamped in the test apparatus to determine the upstream aerosol counts. The agar plates from the cascade impactor are incubated for 48 h and counted to determine the number of viable particles collected. The ratio of the upstream counts to the downstream counts collected for the test specimen are calculated and reported as a percent bacterial filtration efficiency.

Significance and use
This test method offers a procedure for evaluation of medical face mask materials for bacterial filtration efficiency. This test method does not define acceptable levels of bacterial filtration efficiency. Therefore, when using this test method it is necessary to describe the specific condition under which testing is conducted. This test method has been specifically designed

for measuring bacterial filtration efficiency of medical face masks, using *S. aureus* as the challenge organism. The use of *S. aureus* is based on its relevance as a leading cause.

This test method has been designed to introduce a bacterial aerosol challenge to the test specimens at a flow rate of 28.3 l/mm (1 ft³/min). This flow rate is within the range of normal respiration and within the limitations of the cascade impactor. This test method allows the aerosol challenge to be directed through either the face side or liner side of the test specimen, thereby, allowing evaluation of filtration efficiencies which relate to both patient-generated aerosols and wearer-generated aerosols. Degradation by physical, chemical, and thermal stresses could negatively impact the performance of the medical face mask material. The integrity of the material can also be compromised during use by such effects as flexing and abrasion, or by wetting with contaminants such as alcohol and perspiration. Testing without these stresses could lead to a false sense of security. If these conditions are of concern, evaluate the performance of the medical face mask material for bacterial filtration efficiency following an appropriate pre-treatment technique representative of the expected conditions of use. Consider pre-conditioning to assess the impact of storage conditions and shelf life for disposable products, and the effects of laundering and sterilisation for reusable products. If this procedure is used for quality control, perform proper statistical design and analysis of larger data sets. This type of analysis includes, but is not limited to, the number of individual specimens tested, the average bacterial filtration efficiency and standard deviation. Data reported in this way help to establish confidence limits concerning product performance.

5.5.1.2   Test conditions

Inoculum size: 5×10^5 cfu/ml of *S. aureus* (ATCC 6538).
Sample size: 10×10 cm.
Media used: Tryptic soy agar (TSA), Tryptic soy broth and peptone water.
Technique used: Plate count method.
Incubation period: 37°C for 24–48 h.

Test procedure

The bacterial filtration efficiency of the given test sample was performed according to ASTM F2101. The tryptic soy broth (10 ml) was inoculated with *S. aureus* ATCC 6538 and incubated at 37°C for 24 h. The culture was diluted in peptone water to achieve the concentration of 5×10^5 cfu/ml which has challenge delivery rate of 2200 ± 500 viable particles per sample. The challenge suspension was pumped through a nebuliser at controlled flow rate (28.3 Ll/min) and fixed air pressure forming 3.0 ± 0.3 μm aerosol droplets.

The generated aerosol droplets were collected in Anderson's samplers holding six cascade impactor with different pore size. The TSA plates were placed below each cascade impactor. Negative control sampling was done by collecting 2 min sample of air from the aerosol chamber to confirm the sterility. Similarly, positive control was performed with challenge suspension for a minute without test specimen to determine total number of colonies. The test sample was placed above the cascade impactor and the droplets were generated for a minute to check the filtration ability. All the plates were incubated at 37°C for 24–48 h. The colonies were counted and the bacterial filtration efficiency percentage was calculated using the formula given below:

Bacterial filtration efficiency percentage = $C\text{-}T/C \times 100$

Where C = average plate counts for test control

T = average plate count total for test sample.

5.5.1.3   Air permeability

This test method covers the measurement of the air permeability of textile fabrics. This test method applies to most fabrics including woven fabrics, nonwoven fabrics, air bag fabrics, blankets, napped fabrics, knitted fabrics, layered fabrics, and pile fabrics. The fabrics may be untreated, heavily sized, coated, resin- treated, or otherwise treated. The values stated in SI units are to be regarded as the standard. The values stated in inch-pound units may be approximate. This standard does not purport to address all of the safety concerns, if any, associated with its use. It is the responsibility of the user of this standard to establish appropriate safety and health practices and determine the applicability of regulatory limitations prior to use.

Summary

The rate of air flow passing perpendicularly through a known area of fabric is adjusted to obtain a prescribed air pressure differential between the two fabric surfaces. From this rate of air flow, the air permeability of the fabric is determined.

Significance and use

This test method is considered satisfactory for acceptance testing of commercial shipments since current estimates of between-laboratory precision are acceptable, and this test method is used extensively in the trade for acceptance testing. If there are differences of practical significance between reported test results for two laboratories (or more), comparative tests should be performed to determine if there is a statistical bias between them, using competent statistical assistance. As a minimum, ensure the test samples to be used are as homogeneous as possible, are drawn from the material from which the disparate test results were obtained, and are randomly assigned in equal number to each laboratory for testing. The test results from the two laboratories should be compared using a statistical test for unpaired data, at a probability level chosen prior to the testing series. If bias is found, either its cause must be found and corrected, or future test results for that material must be adjusted in consideration of the known bias. Air permeability is an important factor in the performance of such textile materials as gas filters, fabrics for air bags, clothing, mosquito netting, parachutes, sails, tent age, and vacuum cleaners. In filtration, for example, efficiency is directly related to air permeability. Air permeability also can be used to provide an indication of the breathability of weather-resistant and rainproof fabrics, or of coated fabrics in general, and to detect changes during the manufacturing process. Performance specifications, both industrial and military, have been prepared on the basis of air permeability and are used in the purchase of fabrics where permeability is of interest. Construction factors and finishing techniques can have an appreciable effect upon air permeability by causing a change in the length of airflow paths through a fabric. Hot calendaring can be used to flatten fabric components, thus reducing air permeability. Fabrics with different surface textures on either side can have a different air permeability depending upon the direction of air flow. For woven fabric, yarn twist also is important. As twist increases, the circularity and density of the yarn increases, thus reducing the yarn diameter and the cover factor and increasing the air permeability. Yarn crimp and weave influence the shape and area of the interstices between yarns and may permit yarns to extend easily. Such yarn extension would open up the fabric, increase the free area, and increase the air permeability. Increasing yarn twist also may allow the more circular, high-density yarns to be packed closely together in a tightly woven structure with reduced air permeability. For example, a worsted gabardine fabric may have lower air permeability than a woollen hopsacking fabric.

Sampling and test specimens

- Lot sample: As a lot sample for acceptance testing, randomly select the number of rolls or pieces of fabric directed in an applicable material specification or other agreement between the purchaser and the supplier. Consider the rolls or pieces of fabric to be the primary sampling units. In the absence of such an agreement, take the number of fabric rolls or pieces specified. An adequate specification or other agreement between the purchaser and the supplier requires taking into account the variability between rolls or pieces of fabric and between specimens from a swatch from a roll or piece of fabric to provide a sampling plan with a meaningful producer's risk, consumer's risk, acceptable quality level, and limiting quality level.

- Laboratory sample: For acceptance testing, take a swatch extending the width of the fabric and approximately 1 m (1 yd) along the lengthwise direction from each roll or piece in the lot sample. For rolls of fabric, take a sample that will exclude fabric from the outer wrap of the roll or the inner wrap around the core of the roll of fabric.

- Test specimens: From each laboratory sampling unit, take ten specimens unless otherwise agreed upon between purchaser and supplier. Use the cutting die or template described, or if practical, make air permeability tests of a textile fabric without cutting.

- Cutting test specimens: When cutting specimens, cut having dimensions at least equal to the area of the clamping mechanism. Label to maintain specimen identity. Take specimens or position test areas representing a broad distribution across the length and width, preferably along the diagonal of the laboratory sample, and no nearer the edge than one tenth its width unless otherwise agreed upon between the purchaser and supplier. Ensure specimens are free of folds, creases, or wrinkles. Avoid getting oil, water, grease, and so forth, on the specimens when handling.

Preparation of test apparatus and calibration

Set-up procedures for machines from different manufacturers may vary. Prepare and verify calibration of the air permeability tester as directed in the manufacturer's instructions. When using microprocessor automatic data gathering systems, set the appropriate parameters as specified in the manufacturer's instructions. For best results, level the test instrument. Verify calibration for the range and required water pressure differential that is expected for the material to be tested.

Conditioning

Pre-condition the specimens by bringing them to approximate moisture equilibrium in the standard atmosphere for preconditioning textiles as specified in Practice D 1776. After pre-conditioning, bring the test specimens to moisture equilibrium for testing in the standard atmosphere for testing textiles as specified in Practice D 1776 or, if applicable, in the specified atmosphere in which the testing is to be performed. When it is known that the material to be tested is not affected by heat or moisture, pre-conditioning and conditioning is not required when agreed upon in a material specification or contract order.

Procedure

Test the conditioned specimens in the standard atmosphere for testing textiles, which is $21 \pm 1°C$ ($70 \pm 2°F$) and $65 \pm 2\%$ relative humidity, unless otherwise specified in a material specification or contract order. Handle the test specimens carefully to avoid altering the natural state of the material. Place each test specimen onto the test head of the test instrument, and perform the test as specified in the manufacturer's operating instructions. Place coated test specimens with the coated side down (towards low pressure side) to minimise edge leakage. Make tests at the water pressure differential specified in a material specification or contract order. In the absence of a material specification or contract order, use a water pressure differential of 125 Pa (12.7 mm or 0.5 in of water). Read and record the individual test results in SI units as $cm^3/s/cm^2$ and in inch-pound units as $ft^3/min/ft^2$ rounded to three significant digits. For special applications, the total edge leakage underneath and through the test specimen may be measured in a separate test, with the test specimen covered by an airtight cover, and subtracted from the original test result to obtain the effective air permeability. Remove the tested specimen and continue as directed until ten specimens have been tested for each laboratory sampling unit. When a 95% confidence level for results has been agreed upon in a material specification or contract order, fewer test specimens may be sufficient. In any event, the number of tests should be at least four.

Calculation

Air permeability, individual specimens—Calculate the air permeability of individual specimens using values read directly from the test instrument in SI units as $cm^3/s/cm^2$ and in inch-pound units as $ft^3/min/ft^2$, rounded to three significant digits. When calculating air permeability results, follow the manufacturer's instructions as applicable. For air permeability results obtained 600 m (2000 ft) above sea level, correction factors may be required.

Computer-processed data: When data are automatically computer-processed, calculations are generally contained in the associated software. It is recommended that computer-processed data can be verified against known property values and its software described in the report.

5.5.1.4   Capillary flow porometer

Principle

A wetting liquid is allowed to spontaneously fill the pores in the sample and a non-reacting gas is allowed to displace liquid from the pores. The gas pressure and flow rates through wet and dry samples are accurately measured. The gas pressure required to remove liquid from the pores and cause gas to flow is given by

$$D = 4\,\gamma \cos\theta\,/\,p.$$

Where D is the pore diameter,
γ is the surface tension of liquid,

θ is the contact angle of liquid and

p is the differential gas pressure.

From measured gas pressure and flow rates, the pore throat diameters, pore size distribution, and gas permeability are calculated (Figure 5.13).

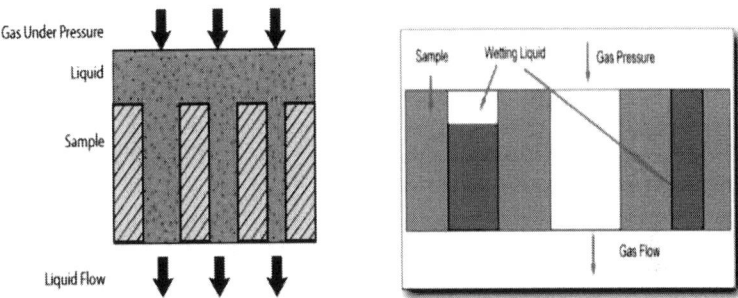

Figure 5.13. Principle of liquid permeability test

The PMI capillary flow porometer is used for R&D and quality control in industries worldwide such as filtration, non-wovens, pharmaceutical, biotechnology, healthcare, household, food, hygienic products, fuel cell, water purification, and battery. Samples often tested include filter media, membranes paper, powders, ceramics, battery separators and health care products.

Application

Advanced capillary flow porometres yield very objective, accurate and reproducible results, considerably reduce test duration, and require minimal operator involvement. Advanced porometers are fully automated and are designed for linear turbulence free test gas flow. The pressure is measured close to the sample and therefore, the correction term in the differential pressure measurement is minimised. Required amount of pressure is uniformly applied on the O-ring seals on the sample and the need for hand tightening the cap on the sample chamber to apply pressure on the O-ring is eliminated. Automatic addition of wetting liquid reduces test time appreciably. This sophisticated instrument has found applications in a wide variety of industries.

Testing capabilities

- Diameter of the most constricted part of a through pore (pore throat)
- Mean flow pore diameter (50% of flow is through pores smaller than the mean flow pore diameter range)
- Pore distribution:

$f = -d[(fw/fd) \times 100]/dD$

fw = flow rate through wet sample.

fd = flow rate through dry sample.

Features

- Liquid permeability: Measuring liquid flow rate through the sample when pressure is applied on excess liquid on the sample. Volume of liquid measured using a penetrometer.
- Pressure hold test
- Hydro-head (break through pressure) test
- Integrity test
- Envelope surface area, average particle size and average fibre diameter obtained from gas flow rate through dry sample
- Multiple sample chamber
- Multiple test mode
- Shuffled smoothness test
- Burst pressure test
- Use of desired fluid including strong chemicals
- Elevated temperature test
- Testing of small samples as well as complete parts
- Any sample geometry (Example: sheets, rods, tubes, hollow fibres, cartridges, and powders)
- Tests in QC, research, or any number of user defined modes
- See-through sample chamber for visual observation of test available
- Real time graphic display
- Window based software for all control, measurement, data collection, data reduction, and report preparation.

Specifications

Pressure accuracy: 0.15% of reading.

Test pressure: 100, 200, and 500 psi instrument versions (700, 1400, 3500 kPa instrument versions)

Pressure and flow resolution: 1/60,000 of full scale (1 part in 60,000).

Maximum pore size detectable: 500 mm.

Minimum pore size detectable:

Flow rates: Up to 200 SLPM (standard litres per minute).

Sample sizes:

Standard: 0.25"–2.5" diameter (up to 1.5" thick).

Standard: 5 mm–60 mm diameter (up to 40 mm thick).

Others available: Sample geometry: Sheets, rods, tubes, hollow Ffibres, cartridges, powders, etc.

Subjective fit test

Fit testing procedures—General requirements.

The employer shall conduct fit testing using the following procedures. The requirements in this appendix apply to all OSHA-accepted fit test methods.

1. The test subject shall be allowed to pick the most acceptable respirator from a sufficient number of respirator models and sizes so that the respirator is acceptable to and correctly fits the user.

2. Prior to the selection process, the test subject shall be shown how to put on a respirator, how it should be positioned on the face, how to set strap tension and how to determine an acceptable fit. A mirror shall be available to assist the subject in evaluating the fit and positioning of the respirator. This instruction may not constitute the subject's formal training on respirator use, because it is only a review.

3. The test subject shall be informed that he/she is being asked to select the respirator that provides the most acceptable fit. Each respirator represents a different size and shape, and if fitted and used properly, will provide adequate protection.

4. The test subject shall be instructed to hold each chosen face piece up to the face and eliminate those that obviously do not give an acceptable fit.

5. The more acceptable face pieces are noted in case the one selected proves unacceptable; the most comfortable mask is donned and worn at least 5 min to assess comfort. If the test subject is not familiar with using a particular respirator, the test subject shall be directed to don the mask several times and to adjust the straps each time to become adept at setting proper tension on the straps.

6. Assessment of comfort shall include a review of the following points with the test subject and allowing the test subject adequate time to determine the comfort of the respirator:

(a) Position of the mask on the nose

(b) Room for eye protection

(c) Room to talk

(d) Position of mask on face and cheeks

7. The following criteria shall be used to help determine the adequacy of the respirator fit:

(a) Chin properly placed;

(b) Adequate strap tension, not overly tightened;

(c) Fit across nose bridge;

(d) Respirator of proper size to span distance from nose to chin;

(e) Tendency of respirator to slip;

(f) Self-observation in mirror to evaluate fit and respirator position.

8. The test subject shall conduct a user seal check, either the negative and positive pressure seal checks. Before conducting the negative and positive pressure checks, the subject shall be told to seat the mask on the face by moving the head from side-to-side and up and down slowly while taking in a few slow deep breaths. Another face piece shall be selected and retested if the test subject fails the user seal check tests.

9. The test shall not be conducted if there is any hair growth between the skin and the face piece sealing surface, such as stubble beard growth, beard, moustache or sideburns which cross the respirator sealing surface. Any type of apparel which interferes with a satisfactory fit shall be altered or removed.

10. If a test subject exhibits difficulty in breathing during the tests, she or he shall be referred to a physician or other licensed health care professional, as appropriate, to determine whether the test subject can wear a respirator while performing her or his duties.

11. If the employee finds the fit of the respirator unacceptable, the test subject shall be given the opportunity to select a different respirator and to be retested.

12. Exercise regimen. Prior to the commencement of the fit test, the test subject shall be given a description of the fit test and the test subject's responsibilities during the test procedure. The description of the process shall include a description of the test exercises that the subject will be performing. The respirator to be tested shall be worn for at least 5 min before the start of the fit test.

13. The fit test shall be performed while the test subject is wearing any applicable safety equipment that may be worn during actual respirator use which could interfere with respirator fit.

14. Test exercises.

(a) Employers must perform the following test exercises for all fit testing methods prescribed in this appendix, except for the controlled negative pressure (CNP) quantitative fit testing protocol and the CNP REDON quantitative fit testing protocol. For these two protocols, employers must ensure that the test subjects (i.e., employees) perform the exercise procedure specified for this appendix for the controlled negative pressure quantitative fit testing protocol, or the exercise procedure of this appendix for the controlled negative pressure REDON quantitative fit-testing protocol. For the remaining fit testing methods, employers must ensure that employees perform the test exercises in the appropriate test environment in the following manner:

(1) Normal breathing. In a normal standing position, without talking, the subject shall breathe normally.

(2) Deep breathing. In a normal standing position, the subject shall breathe slowly and deeply, taking caution so as not to hyperventilate.

(3) Turning head side to side. Standing in place, the subject shall slowly turn his/her head from side to side between the extreme positions on each side. The head shall be held at each extreme momentarily so the subject can inhale at each side.

(4) Moving head up and down. Standing in place, the subject shall slowly move his/her head up and down. The subject shall be instructed to inhale in the up position (i.e., when looking toward the ceiling).

(5) Talking. The subject shall talk out loud slowly and loud enough so as to be heard clearly by the test conductor. The subject can read from a prepared text such as the Rainbow Passage, count backward from 100, or recite a memorised poem or song.

(6) Grimace. The test subject shall grimace by smiling or frowning.

(7) Bending over. The test subject shall bend at the waist as if he/she were to touch his/her toes. Jogging in place shall be substituted for this exercise in those test environment units that do not permit bending over at the waist.

(8) Normal breathing.

Each test exercise shall be performed for one minute except for the grimace exercise which shall be performed for 15 s. The test subject shall be questioned by the test conductor regarding the comfort of the respirator upon completion of the protocol. If it has become unacceptable, another model of respirator shall be tried. The respirator shall not be adjusted once the fit test exercises begin. Any adjustment voids the test, and the fit test must be repeated.

5.6   Results and discussions

5.6.1   Test results of bacterial filtration efficiency

Standard: ASTM F 2101

Bacterial filtration efficiency (BFE) $\% = C - T/C \times 100$,

where C = average plate count for test control and T = average plate count total for test sample.

Observation:

S. No.	Test particulars	Result
1	Sample	Zinc oxide treated fabric
2	Area of test specimen	Ø 100 mm

3	Flow rate of aerosol	28.3 l/min			
4	Mean particle size of challenging aerosol	3.3 μ			
5	Average plate count of positive control	1740.2			
6	Average plate count of negative control	0			
7	Plate count total for each stage	Plate no	Positive control	Trial 1	Trial 2	Trial 3
		1.	4565	318	321	319
		2.	3851	182	186	178
		3.	978	123	136	128
		4.	537	95	109	98
		5.	324	22	19	23
		6.	186	0	0	0
		Total	10441	740	771	746
		Average	1740.2	123.3	128.5	124.3
8	BFE of test specimen	Trial 1		Trial2		Trial3
		92.9%		92.6%		92.8%
9	Average BFE	**92.8%**				

BFE: Bacterial filtration efficiency.

Standard acceptance level for surgical mask is 95% and comparatively for the developed anti-pollution mask it is 92.8% and hence it is sufficient to filter the air borne particles during peak pollution prevalence according to standard norms. And most of the commercial products focus on filtering the particulate matters that the mask don't filter out or kill microbes hence our developed product gives an edge over the other available products comparatively at low cost.

Normally, the use of polypropylene gives an advantage over the developed product since it has anti-fungal properties.

5.6.2   Test results for breathability test

1. Breathing resistance
   - At 95 LPM (litres per minute) with one layer–19 Pa/cm^2

- The differential pressure should ne <29 or < 49 Pa/cm^2 is required for surgical mask according to EN 14683 standard.
- The differential pressure is an indicator of the breathability of the mask expressed in differential pressure in Pa/cm^2.
- Hence, the differential pressure of the developed mask is suited according to the standard norms and it ensures good breathability.
- The main drawback in most of the existing commercial product is the breathability and our product therefore ensures good breathability

2. Air permeability
- Standard: ASTM D 737:04
- Test pressure—125 Pa and test area–38 cm^2
- Sample testing condition: R.H. 65% ± 2% and temperature 21°C ± 1°C
- Air permeability of the developed mask is 117 cc/cm^2/sec.
- The air permeability decreases with increase in fabric weight. The resultant developed product weight is less than 200 GSM hence it has good air permeability. This is because increase in fabric weight causes more number of fibres per area and causes the fabric density to increase the resistance to air flow.
- With the test it can be concluded that this fabric attains maximum collection efficiency at a shorter interval of time hence it falls under acceptance category based on the norms.
- The air permeability is higher in accordance to the hollow structure of the fibre since the density is less. Therefore it reaches higher permeability at pressure drop

5.6.3   Test results for capillary flow porometer

- Test method—capillary flow porometer
- Type of liquid—gal wick
- Type of method—wet-up and dry-up
- Pressure flow at smallest pore—0.547 PSI
- Pressure flow at largest pore—0.028 PSI
- Bubble point pressure—0.028 PSI
- 10% cumulative filter flow occurs at 163.3964 µ
- 25% cumulative filter flow occurs at 130.2915 µ
- 75 % cumulative filter flow occurs at 65.7175 µ

- 90% cumulative filter flow occurs at 40.5085 μ
- Filter flow % = 100 × wet-flow / dry-flow
- Average filter flow % = 98.07
- At 100 LPM (litres per minute) with three layer it is 0.6 PSI.

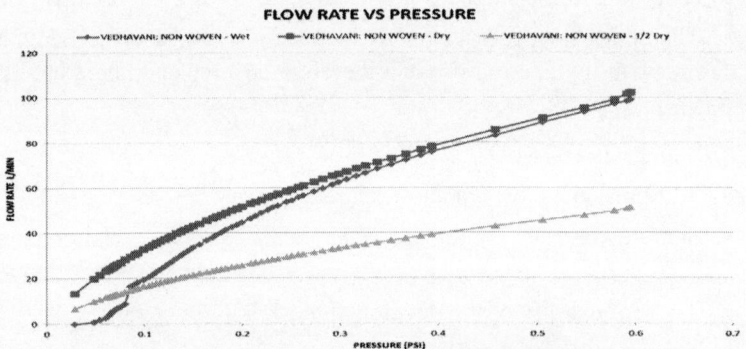

Figure 5.14. Graph between flow rate and pressure

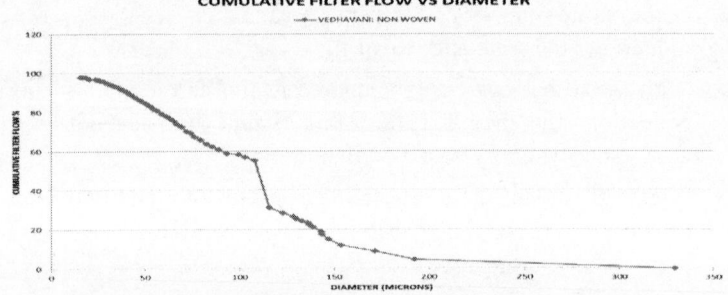

Figure 5.15. Graph between filter flow and diameter

- The filtration efficiency increases and after sometimes it almost reaches a steady state. The increase of efficiency with time and pressure can be explained that as the particles reach the fabric along with air. So the deposition of particles occurs and filtration takes place (Figures 5.14 and 5.15).
- The collection efficiency for hollow filter fabric attains maximum % at shorter interval of time.
- The products density is quite high hence the resistance to the flow of particles and easily separates dust particles from air stream.
- The granular activated carbon filter fabric has the ability to resist and withstand particles up to less than 2.5 μ and hence filtration efficiency of the product overall is high.

5.6.4   Subjective fit and product test analysis

SUBJECTIVE PRODUCT TEST ANALYSIS

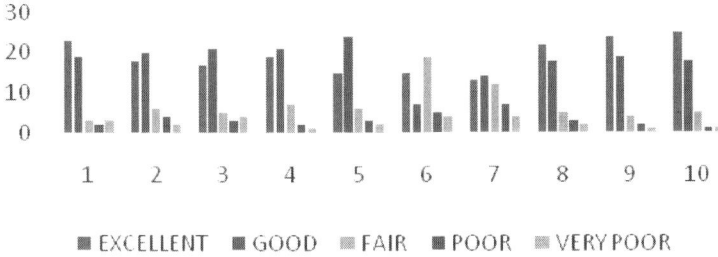

■ EXCELLENT   ■ GOOD   ▓ FAIR   ■ POOR   ▓ VERY POOR

Figure 5.16. Graph for subjective fit analysis

Based on the survey results, we can conclude that the overall performance of the developed products falls under excellent, good and fair. And it is acceptable by the customers (Figure 5.16).

The fit of the product seems to be good according to the survey and customers prefer it since its fashionable and reasonable price.

5.6.5   Comparison

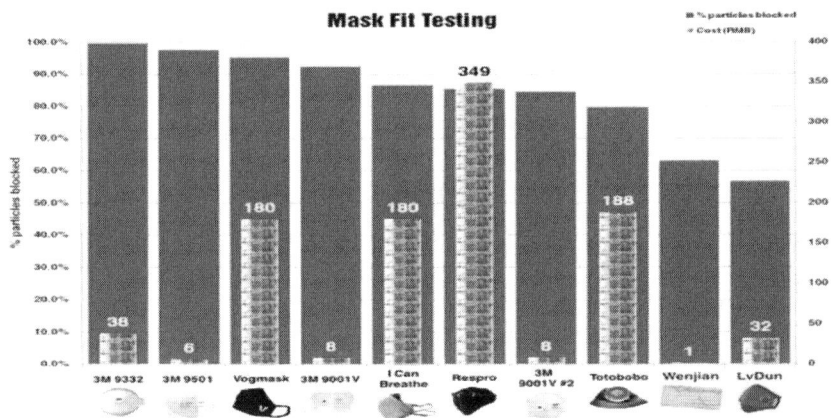

Figure 5.17. Mask fit testing

This chart explains the commercially available products percentage of particles blockage and their respective prices in the market (Figure 5.17).

In comparison with this chart, the developed anti-pollution mask has about more than 95% of filtration efficiency according to norms and it equals to the high quality product.

In terms of cost of the best product, the developed mask is relatively cheaper and affordable at the same time it is proved to be efficient.

5.7   Conclusion

The bacterial filtration efficiency of the developed product is acceptable by the standard norms by 92.8% in comparison with the surgical mask. And many of the commercially available product does not focus on the bacterial filtration efficiency hence this feature of the product gives an added advantage in market.

Since cotton is the innermost layer which is in contact with the skin, probably no allergic reactions or irritation is caused to the wearer.

The filtration efficiency of the granular activated carbon is high and has the ability to filter out the particulate matter of less than 2.5 μ and NOX and SOX compounds.

The filter flow percentage of the tri-layered fabric is 98.07%.

The air permeability is acceptable according to the standard norms by 117 cc/cm²/sec.

The pressure drop of the fabric is around 20 Pa which falls under the acceptance level.

The price of the product is reasonable and affordable compared to commercially available products and also gives a high filtration efficiency.

Hence, the product is fashionable, reasonable and affordable with high performance filtration.

5.8   References

1. Anon, (2013), The American heritage dictionary of the English language, 4th edn, Houghton Mifflin Harcourt company, retrieved October 15th from http//images.yourdictionairy.com/.
2. ASTM F 2100, (2007), Standard specification for performance of materials used in medical face masks, USA, 390–392.
3. Belkin, N.L., (2009). The Surgical mask has its first performance standard, a century after it was introduced, Bull Amer College Surgeons, 94(12): 22–25.
4. Hamilton, C.D., (1915), The effect of typhoid vaccination on the widal reaction, Journal American Medical Association, 95(22): 1873.
5. Hayavadana, J. and Vanitha, M., (2009), The world of surgical textiles surgical mask, Asian Textile Journal, 18(12): 33–35.
6. Lunenschloss, J., and Albrecht, W., (1985), Nonwoven bonded fabric. Ellis Horwood Limited, 396–397.
7. McCarthy, B.J., (2011), Textiles for hygiene and infection control, Woodhead Publishing Ltd, 125–135.

8. Pal, S., Tak, Y.K., and Song, J.M., (2007), A study of the gram-negative bacterium *Escherichia coli.* Applied Environmental Microbiology 73(6): 1712–1720. Doi: 10.1128/AEM.02218-06.

9. Ashe, B., (2011), A detailed investigation to observe the effect of zinc oxide and silver nanoparticles in biological system, National Institute of Technology.

10. Piekaar, H.W., and Clarenburg. L.A., (1967), Aerosol filters – pore size distribution in fibrous filters. Chemical Engineering Science, 22: 1399–1407.

11. Rawal, A., Lomov, S., Ngo, T., and Vankerrcbrouck, J., (2007), Mechanical behavior of thru-air bonded nonwoven structures, Textile Research Journal, 77: 417–431.

12. Rollin, A.L., Denis, R., Estaque, L., and Masounave, J., (1982), Hydraulic behavior of synthetic nonwoven filter fabrics, The Canadian Journal of Chemical Engineering, 60: 226–234.

13. Simmonds, G.E., Bomberger, J.D., and Bryncr, M.A., (2007), Designing nonwovens to meet pore size distributions, Journal of Engineered Fibres and Fabrics, 2: 1–15.

14. Unit of Concerned Scientists.

6

Eco-management in apparel industry

Dr. M. Parthiban and Dr. M. R. Srikrishnan

Department of Fashion Technology, PSG College of Technology,
Coimbatore – 641 004
Email: parthi111180@gmail.com

In India, each state has its own pollution control authority. This authority mainly deals with water pollution by textile industry. The aim is to ensure that the effluent water being discharged into city sewage, stream, river or sea is not harmful to human, animal or plant life. In order to get the parameters of effluent water to suitable standards, the effluents are treated by effluent treatment plant. It can be stated that basically no steps are taken by pollution control authorities to control air and noise pollution in textile industry. In the case of toxicity of textile products, the awareness is increasing in India due to rigid rules and regulations being set up by developed countries. It has forced Indian producers to fulfill these rules and regulations for attracting exporters. In eco-management systems followed in textile industry, water utilized for washing is re-used.

Keywords: Ecology, ISO & Environment, Eco-Standards, Eco-label, Effluent Treatment & EMS

6.1   Introduction

The present Indian domestic market size is approx. US \$350 billion by 2024–25. Indian exports of textile and apparel products leads to a great income of the country. Production normally involves some kind of disposal. Not all the disposal is eco-friendly and similarly production. This eco-friendly procedure for production and disposal involves a higher end cost rather than the regular production. But this eco-friendly procedure for production and disposal of textile industry leads to satisfy longer needs than the other. Hence, in order to satisfy longer needs and also be financially manageable, giving a handsome of profit to the producer management is necessary. Thus this report discusses about the eco-management in textile technology.

6.2   Ecological consideration

- Production ecology
- User ecology
- Disposal ecology.

6.2.1   Production ecology

Production ecology mainly involves the reduction of pollution caused by the textiles and apparel production. The pollution mainly involved here will be air pollution and water pollution. Air pollution mainly will be due to the mills that leak out harmful, carcinogenic gases into the air during production. Water pollution is the effluents mixed with water due to unknown or known knowledge of the producer during dying and other process of textiles.[1]

6.2.2   User ecology

User ecology primarily involves in the usage of different kind of materials properly, effectively and in a proper way. This user ecology primarily concerns with the consumers. The mind-set of the consumers decides about the user ecology. The main activity to be done to reduce the ill effects caused by consumers is to reduce the pollution and unnecessary waste caused by them. This also involves the other mills and industry who use the textile garments and apparels for their purpose and working with that in an effective way.

6.2.3   Disposal ecology

Disposal ecology concerns with both producers and consumers. Producers who manufacture textile and apparels must take care that the waste cloth is either recycled or used in a proper way again and not just disposed so that the disposal becomes in an effective and eco-friendly manner. With respect to the consumers, every consumer must use the textile and apparel to their maximum so that the waste generated becomes less and the land pollution caused by it is further reduced giving eco-management in a simple way.[2]

6.3   Goals and strategies for eco-management

6.3.1   ISO and the environment (ISO 14001:2004)

ISO 14001:2004 sets out the criteria for an environmental management system and can be certified to. It does not state requirements for environmental performance but maps out a framework that a company or organisation can follow to setup an effective environmental management system. It can be used by an organisation regardless of the activity or sector. Using ISO 14001:2004 provides assurance to company management and employees as well as external stakeholders that environmental impact is being measured and improved.[3]

6.3.2   Environmental management system

- Use of eco-friendly raw material
- Same procedure for finishing and packing
- Textile industry needs to recognise this and take necessary steps to switch over safer alternatives whenever possible
- Adoption of ISO 14000 environmental management system help to meet requirements of regulatory authorities
- Effective tool in enhancing market share
- Optimum utilisation of resources like water and electricity
- Enhance and ensure the sustainable development and market acceptability.

6.3.3   Eco-standards and eco-labels

The concept of eco-friendly textiles is promoted by eco-standards and eco-labels. Eco-labels are based on environmental friendly norms for various chemical stipulated on the basis of cradle to grave approach. Use of eco-labels is voluntary in nature.

Chemicals considered for eco-norms

- Formaldehyde
- Toxic pesticides
- Pentachloro phenol
- Heavy metal traces
- Carcinogenic azo dyes
- Halogen carriers
- Chlorine bleaching.

6.3.1   Effluent characteristics

Textile effluents are generally classified into

- Coloured
- High Biological Oxygen Demand (BOD)
- High TDS Total Dissolved Solids (TDS)
- High ratio between BOD and Chemical Oxygen Demand (COD)
- Ratio of 1:2 to 1:3 indicate good bio-degradability
- In wool ratio 1:5 indicating difficult bio-degradability due to grease context.

6.4   Statistics

One of the oldest and single large industrial sectors is the textile industry. Yet many problems which were there while starting are the same as of now. Statistics says that the organised sector in textile industries is less than 75%. It also states that the disorganised sector and decentralised power loom is less than 25%. And it also states that the organised sector presently is 6%.[4,5]

6.5   Reforms to be done

6.5.1   Scale and investments

- Maintaining competitive exchange rates
- 15% investment allowance
- Hire–purchase of looms and knitting
- Goods and service tax (GST) implementation
- Service tax exemption
- Equity fund for start-ups and expansion
- Mega textile parks
- Attracting FDI.

6.5.2   Skill, quality and productivity

- Scaling up existing skill development initiatives
- Tax relief on skill development agency
- National manufacturing competitiveness programme
- Worker housing to be an intrinsic part of textile parks
- Promoting machinery manufacturing.

6.5.3   Labour law reforms

- Avoiding women night shifts
- Fixed term employments
- Reviewing overtime caps
- Industrial dispute act norms
- Contract labours to be allowed.

6.5.4   Focus on value addition

- Plug and play garment sector
- Export finance at 7%
- Promotion of technical textile sector.

6.6   Results of eco-management in textile industry

6.6.1   Supply chain goal

- Water reduction
- Time compression
- Flexible response
- Unit cost reduction.

6.6.2   Financial benefits

- Higher profit margins
- Improving cash flow
- Revenue growth and higher return on assets.

6.6.3   Supply chain improvement

- Product/process innovation
- Threats/substitutes
- Partnership.

6.6.4   End customer benefits

- Improved product quality
- Improved timeliness
- Improved value
- Improved flexibility.

6.7   Conclusion

Thus, the eco-management of textile industry could be achieved in several ways so that it results in a great level improvement to the society financially as well as socially. Understanding the very much need of eco-management in textile industry could be a great way to long time sustainability in the market. Thus, this eco-management in textile industry could be a great way for success in many ways.

6.8   References

1. Winkler T, Muhlendahl C V and Fischer T, "Green Technology in Textile Processing", Int. Tex. Bull, 44 (1998), 35.

2. Edward Menzes, and Dr Bharat Desai (Rossari Biotech): Water in Textile Wet Processing- Quality and Measures, Clothesline, 2004.
3. Vigneswaran C & Keerthivasan D, "Eco-standards in garment processing through eco-auditing", The Indian Textile Journal, 2013.
4. Hillary, R. (2004). Environmental management systems and the smaller enterprise, Journal of Cleaner Production, Vol. 12, Issue 6, pp. 561-569.
5. http://www.cleanclothes.org/resources/publications/fatal-fashion.pdf/June 27, 2013.

7

3P's concepts for sustainable development

Dr. M. Parthiban and Dr. P. Kandhavadivu

Department of Fashion Technology, PSG College of Technology,
Coimbatore-641 004
Email:parthi111180@gmail.com

The term "3P" refers to a business model developed to encourage social responsibility and sustainability among businesses worldwide. The corporations who adopt these standards are known as "triple bottom line," or TBL, companies. This term is attributed to John Elkington, founder of the consulting firm Sustainability, and author of "Cannibals with Forks: the Triple Bottom Line of 21st Century Business." The three Ps stand for "people, planet and profit."

7.1   Introduction

A sustainable business works towards achieving a positive triple bottom line—people, planet, profit—so that it generates profit whilst minimising any negative impacts to people and our planet through its products, services and operations.[1]

- **People:** All individuals are treated fairly. No group is harmed, exploited, or unequally burdened by business pursuits.
- **Planet:** The Earth's natural resources (including the ecology, plants, or wildlife species) are not adversely impacted.[2]
- **Profit:** Fiscal or economic successes are not limited or unattainable by the pursuit of the other two values.

This concept is especially relevant in the growing global markets and developing industries in today's society.

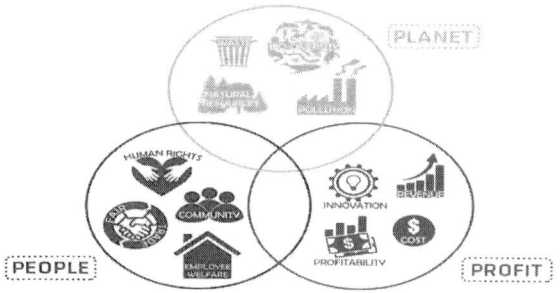

7.2   Necessity for 3P concepts

- Second most polluting industry in the world—affecting eco-system.
- Building more efficient operations will lower your costs, increasing your profits and helping you to drive growth and innovation.
- Gaining competitive advantage, improving brand value and reputation, and most importantly, decrease your environmental footprint.[3]

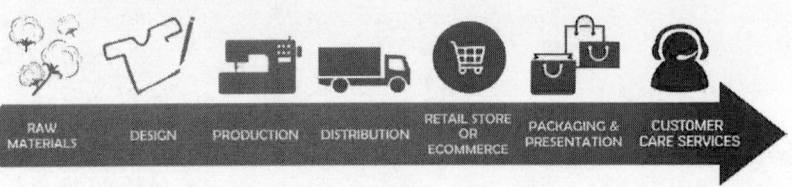

7.3   Analyse your operations—Key

- Analysing your full operations is key to getting going with your sustainable business.
- All the way from raw materials through to your final product and beyond, you will be making decisions for your brand that will affect its footprint.
- Analysis in all the stages are necessary and compulsory!
- The first step—SWOT (strength, weakness, opportunity, threat) analysis which will help you determine your capabilities in the global marketplace.
- Analysing the environmental and social impacts of your business through your entire supply chain and operations.[4]
- This will allow you to make a plan to tackle issues, with the quick wins first through to the issues that may need longer investment and research.
- Start with achievable targets and don't get overwhelmed. Pick the impact areas that matter most to your business.
- For example, look for stages in your supply chain where you can purchase sustainable materials, reduce waste, save energy, streamline production and decrease transport emissions and you will be taking the first steps to creating your sustainable fashion business![4,5]

7.4   Raw materials

- Selection of raw materials that are more environmentally sustainable.

- For example, use of recycled materials/organic cotton/new dyeing technologies that require less water and chemicals; or materials that have high recyclability at the end of the product's lifecycle.
- Purchase raw materials in bulk.

7.5   Design

- After choosing your materials, look at how to reduce your environmental impact through design.
- Could you use design techniques to reduce waste? Could you design clothes with multiple uses and functions?
- Could you improve the product's lifecycle through design?
- Could you foster good consumer use, such as wash, care, repair, reuse and recycle through your design?

7.6   Production

- As you move into production, you can think about how you can reduce energy and water consumption in your production process and factory operations.
- Could you reduce wastage of raw materials?
- Are you aware of the use of hazardous chemicals in your production?
- Could you monitor your production line to comply with regional and international environmental and safety requirements?
- Could you avoid over-production by producing just the right amount according to your demand?

7.7   Distribution

- How you distribute your products can also be more environmentally-sustainable so that you reduce pollution and greenhouse gas emissions during transpiration and handling?
- How about shipping or road or rail transportation instead of airfreight?
- Could you choose more energy-efficient modes of vehicle?
- Could you minimise your packaging to reduce weight?
- Could you use a more efficient packing system when stacking goods to reduce bulk?
- Is the space in your storage or distribution centre well used?
- Could you move your production closer to your operations and distribution?

7.8   Retailing

- How you ultimately make your sale to your customer, whether through physical retail or an online platform, also affects the environment?
- Physical retail spaces consume a significant amount of energy, through electricity bills and through all of the necessary visual merchandising that is needed for your products.
- Simple steps to alleviate this are to opt for a green energy supplier and select energy-efficient lighting and to seek more sustainable sources for your visual merchandising and carrier bags.
- Do your research for sustainable options in these areas?
- Selling online, otherwise known as e-commerce, might shift your environmental footprint from one aspect of your business to another by negating the needs of the physical shop.
- For example, if you decide to be a sole e-commerce company then the majority of your impact would likely be related to packaging and customer delivery. To work towards negating this, you could partner with a more energy-efficient and low carbon logistics partner.

7.9   Packaging and presentation

- How you decide to present and package your goods can also leave a mark on the earth. How do you package your products for shipping?
- What material is your packaging made of?
- Is the packaging recyclable?
- Have you communicated your sustainable values on packaging to your merchandisers or the manufacturers who pack your products?
- You can also think about the materials you use to present your products at the point of purchase.
- Could you minimise the amount of printing for visual merchandising?
- Could you use more environmentally sustainable options for hangers, hang tags and packaging?
- Could you re-use any of the materials?

7.10   Customer care services

- In recent years, responsible fashion brands and retailers have increasingly been focusing more attention on educating their customers about better consumer care by offering improved care labels and instructions and by adding new services, like take-back services, repair and garment rental, to enhance the sustainability of products.

- You can also offer customer service options that extend the life of your garments, divert waste from landfills and build brand loyalty.
- Need to know how your products are being used, cared for and disposed of in order to make more informed decisions during the design stage.
- Could you design your products to influence how they are being washed, cared, stored, re-used and disposed?
- Could you create clothes that make repairs easier?
- Could you design clothes with their end-of-life in mind?

7.11   Value sharing

- A company that has strong sustainable values holds a stronger brand identity. By building a sustainable business, you will attract employees with similar visions of positive environmental impact along with a loyal consumer base. So it is important to engage with your stakeholders.
- Are your suppliers and employees clear of the sustainability of your brand?
- Do you communicate your sustainable story to your customers and buyers?
- Have you thought of running activities that include and benefit the wider community?

7.12   Reporting to stakeholders

- Corporate social responsibility (CSR) is the conventional name for the commitments by a business to behave ethically and to contribute positively to society and to the environment.
- A CSR team is responsible for implementing strategies and communicating them to stakeholders.
- Over time, the role of CSR has been evolving from a periphery department to becoming more integrated into all business aspects.
- Along the way, the name of this function has also changed and it is more often now known simply as a sustainability team.
- However large your business, reporting is a key part of CSR or sustainability.
- And now more than ever, as the pace of information flow is accelerated and customers demand transparency of the brands they buy, it's important to show accountability for your business' impacts to your stakeholders.
- Your sustainability targets come into play here as a key measurement of how you are doing on your journey.

7.13   Customer engagement

- It's important to engage your customers in your sustainability initiatives as they will be your biggest advocates.
- Share your sustainability efforts with your customers as part of your marketing strategy and build trust by connecting with their values and motivations.
- Need to adopt a proactive and transparent strategy to let your customers know about the positive ways your business has engaged in sustainable practices.
- But also be honest about the challenges that you are still working to improve.

7.14   Employee engagement

- As well as communicating with your customers, holding shared values within your organisation is also very important.
- Your team is the backbone of your business, so engaging them on the journey to sustainable business can have multiple benefits.
- Communicate your core values to your whole company, even your junior employees, and ask for their ideas and inputs for improvements.
- Research shows that when employee and company values match, employee loyalty, creativity, quality and accuracy of communication and integrity of decision-making increase!
- Allocating sustainability champions to help you address and succeed in the major challenges will further drive their engagement.

7.15   Community engagement

- Engaging with the wider community is also a useful step that can lead to long-term gains.
- Working with people beyond your customers and employees will help you build brand reputation and spread the message about your work.
- You may also discover ways in which the wider community can help you achieve your sustainability targets.
- You might want to engage people in the local community where your company operates, e.g., your suppliers, local schools, charities and NGOs or even your local authority.
- And don't forget the media too!

7.16   Conclusion

The truth is, the methodology itself has been around for a long time now. It exists in every successful business, industry, niche, system, etc. Any successful business owner can likely identify this process in their own business, even if they've never put their management practises into this context. If you have a failed business or two under your belt, the 3P's could have saved you. If you're starting a new business, or need to revamp a struggling one, the above information should be invaluable to you. The 3P's always contain much of the same elements; they're just interchanged and moved around to suit the established process of a given industry or individual business. Sometimes a fourth "P" is also added for "Performance". It's all subjective, but the core ideas are always the same nonetheless. So enough of the digressions, let's get on with how "People", "Planet", and "Profit" can help you create a successful business model and/or revamp existing ones that need a little kick in the pants to get business flowing.

7.17   References

1. http://www.entrepreneur.com/article/80678/Jan 10, 2018
2. https://www.americanexpress.com/us/small-business/openforum/articles/4-ways-to-create-a-product-that-sells-itself/Feb 27,2018
3. http://www.thelaunchcoach.com/product-launch-tips-3/May 28,2014
4. http://www.forbes.com/sites/georgebradt/2011/05/11/follow-three-imperatives-in-starting-a-successful-service-business/May 28, 2014
5. http://techcrunch.com/2011/04/24/what-should-you-do-with-your-crappy-little-services-business/mar 28,2014

8
Enzymatic approach to sustainable apparel production and green chemistry

Gopalakrishnan D,[1] Apoorva Gupta[2] and Dr. (Mrs.) Meenu Srivastava[3]

[1] Department of Fashion Technology, PSG College of Technology,
Coimbatore – 641 004
Email: dgk.psgtech@gmail.com
[2,3] Department of Textiles and Apparel Designing, College of Home Science,
Maharana Pratap University of Agriculture and Technology, Udaipur,
Rajasthan – 313 001
Email: [2] apoorvafsn22@gmail.com and [3] meenuclt@yahoo.com

Abstract: Consumers are increasingly demanding more natural and greener products. Although it's difficult to find a single definition, natural refers to the source of the raw materials, while green refers to the process used to convert starting materials into a finished ingredient. Today, buyers not only care about the ingredients inside their favourite products, but they also care about the impact of manufacturing those products has on the environment. Many consumers prefer to purchase environmentally friendly products, especially for their personal care and home needs. Enzyme is one of the oldest technologies, is becoming a favourable alternative to chemical processes and a vital part of green technology. It offers green and clean solutions to chemical processes and is emerging as a challenging and revered alternative to chemical technology. The chemical processes are now carried out biologically by biocatalysts which are integral components of any biological system. However, the utility of enzymes is not native to us, as they have been an integral part of our lives from immemorial times. Commercially available enzymes are derived from plants, animals, and microorganisms. However, a major fraction of commercially available enzymes are derived from microbes due to their ease of growth, nutritional requirements, and downstream processing. To meet the increasing demand of robust, high turnover, economical and easily available biocatalysts, research is always channelized for novelty in enzyme or its source or for improvement of existing enzymes by engineering at gene and protein level.

Textile processing has benefited greatly on both environmental and product quality aspects through the use of enzymes. The application of enzymes in diverse textile industries indicates a positive trend which needs to be satisfied with the discovery of novel enzymes and metabolites. Green chemistry provides a technical solution to many environmental problems. It is effective due to design stage efforts, starting at the molecular level lets one to design out the hazardous properties and to design in environmentally appropriate features.

Key words: enzymes, green chemistry, biocatalysts, green and clean, microorganisms

8.1   Introduction

The global market is expected to experience a growth of 6% in enzyme requirement with an estimated market of $7 billion in 2013. North America and Western Europe are predicted to show an increased growth, while the highest growth is likely in developing countries of Asia, Africa and Mideast regions, along with Latin America and Eastern Europe. China is emerging as an important base and market for industrial enzymes due to various R&D activities set-up by many industrial giants which accounts for 10% of the global scenario. The demand of diagnostic and therapeutic enzymes is expected to increase owing to improvement in medical care facilities in developing countries and global healthcare reforms. Few enzyme manufacturing industries in the world include AB Enzymes GmbH, Advanced Enzyme Technologies Ltd., Amano Enzyme Inc., Asahi Kasei Pharma Corporation, Cargill Texturizing Solutions, Genencor International Inc., DSM Food Specialties, Hayashibara Company, Nexgen Biotechnologies Inc., Novozymes A/S, and Maps Enzymes Ltd. The global industrial enzyme market has evolved continuously due to numerous mergers and acquisitions. In the year 2011, enzyme industry giants like Novozymes and DuPont occupied market shares of 47% and 21%, respectively. Technical enzymes were valued at $1.2 billion in 2011, and this is expected to rise to $2.2 billion in 2016 with the highest sales predicted in the leather and bioethanol markets. Similarly, food and beverage enzyme sector is expected to achieve about $2.1 billion by 2016, from a value of $1.3 billion in 2011 as shown in Figure 8.1. This is well correlated with the numerous patents which have been filed over a period of years which indicate an increasing trend. From Figure 8.1, it can be inferred that due to the lack of information on intellectual property rights (IPRs) in the 1970s, there were hardly any patents on any of the industrial enzymes. But from the year 2000 till date, there has been a tremendous increase in the number of patents filed or obtained for various enzymes. The maximum number of patents is for proteases followed by amylases and cellulase perhaps due to maximum utility of these enzymes. The application and issuing of patents for various enzymes is expected to grow in the future due to green technologies.

These specialty enzymes also include those useful in medicine and biotechnology, e.g., kinases, polymerases, and nucleases. Besides this, their utility in wide range of personal care products is revolutionizsing the cosmetic industry too. The specialty sector is expected to reach $4.3 billion by 2015, and industrial enzyme segment is worth $80 million, according to reports by reputed market researchers and industry analysts.

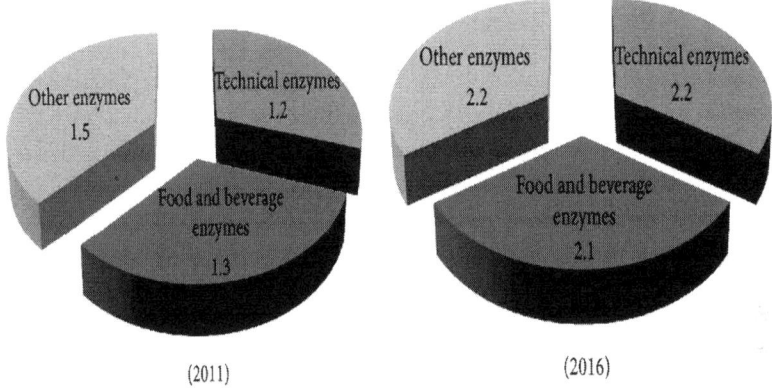

Figure 8.1. Application of specialty enzymes.
Courtesy: Divya Prakash et al., (2013).

8.2   Importance of biotechnology

Biotechnology, in pure scientific term, is defined as "Application of biological organisms, systems and processes to manufacturing and processing industries". Defining the scope of biotechnology is not easy because it overlaps with so many industries, such as the chemical industry and food industry being majors, but biotechnology has found many applications in textile industry also, especially in genetic engineering, textile processing and effluent management (Emilia Csiszar et al. 1998; Traore and Buschle-Diller 2000). It is important to note that biotechnology is not just concerned with biology, but it is a truly interdisciplinary subject involving the integration of natural and engineering sciences (Tyndall 1996). A bio-process is any process that uses complete living cells or their components (e.g., bacteria, enzymes, chloroplasts) to obtain desired products. It also offers the potential for new industrial processes that requires less energy and are based on renewable raw materials (Gubitz and Cavaco-Paulo 2001; Opwis et al. 2006).

8.3   Concept of green chemistry

Green chemistry looks at pollution prevention on the molecular scale and is an extremely important area of chemistry due to the importance of chemistry in our world today and the implications it can show on our environment. The green chemistry program supports the invention of more environmentally friendly chemical processes which reduce or even eliminate the generation of hazardous substances.

- Green chemistry is the chemistry that
- Doesn't hurt nature,
- Reduce or eliminate the use or generation of hazardous substances,
- Provides more eco-friendly alternative,
- Prevents formation of waste,
- Creates new knowledge based on sustainability, i.e., sustainable chemistry,
- Takes a life cycle approach to reduce the potential risks throughout the production process. Life cycle analyses (LCAs) enable a manufacturer to quantify how much energy and raw materials are used, and how much solid, liquid and gaseous waste is generated, at each stage of the product's life.

8.3.1   Principles of green chemistry

Twelve principles of green chemistry are as follows:

1. Prevent waste: Design chemical syntheses to prevent waste, thereby eliminate/minimise waste treatment processes. It is better to prevent waste than to treat or clean-up waste after it is formed.
2. Maintain atom economy: There should be few, if any, wasted atoms.
3. Use safe chemical synthesis methods.
4. Use low toxic products: Use fully effective but safe or non-toxic chemicals and products.
5. Choice energy efficient processes: Prefer ambient temperature and pressure reactions.
6. Use renewable feedstock: Use non-depleting renewable agricultural products or the wastes of other processes and not products derived from fossil fuels.

7. Omit derivation steps: Follow least number of sequential chemical steps, and choose direct reactions.
8. Catalysis: Catalytic reactions generate minimum waste—its little amount can carry out a single reaction many times. Gold is an outstanding catalyst for oxidation processes.
9. Safer solvents and auxiliaries: Use aqueous or other safe media.
10. Degradation of chemical products: Choose chemicals degradable to harmless substances.
11. Real time analysis: Minimise/eliminate by-products by real-time monitoring and control.
12. Safety: Assure minimum chemical accidents (e.g., explosions, fires and harmful releases).

8.3.2   Assessment of green reactions

The easiest way to assess how green a chemical process is to measure the amount of waste generated. E-factor (environmental acceptability) measures the ratio of the mass of waste to that of the product. All processes should aim for the lowest possible E-factor—for truly green processes, the E-factor should be zero. Large-scale manufacturing units for bulk chemicals may generate large amount of waste, but their E-factors may be smaller than those of small-scale units as the E-factor depends on the quantity of waste in relation to total production.

8.3.3   Misconceptions

Some misconceptions about green chemistry are as follows:

- **Cost benefit:** Green chemistry truly allows for increased profits by saving reagents, solvents, energy, waste disposal costs, personnel costs, and increasing production.
- **Perfection of the systems:** A perfectly green process may not be so green if it hasn't been applied in the right situations.
- **Fields of application:** The application of green chemistry is not restricted. It is applicable to various industrial sectors since all industrial processes involve one or more of the following basics: raw materials, chemical reactions, solvents, and separation/purifications.
- **Longevity:** Green chemistry often remains unchanged for long periods of time.
- **Overall performance:** Traditional purification and separation methods both generate large amounts of acid, base, and solvent wastes, and are often energy intensive. New separation techniques

such as carbon dioxide extraction, phase separation, evaporation, membrane separation, and reforming by-products into new products minimise waste generation.

8.4   Textile industry and pollution

The textile industry is considered as the most ecologically harmful industry in the world. The utilisation of rayon for clothing affects fast depleting forests. Petroleum-based synthetic fibre and the dyes are not sustainable and not biodegradable. Cotton cultivation requires large quantities of synthetic fertilizers and pesticides/herbicides. Presently the conventional cotton crops occupy 3% of the world cultivated areas. Nevertheless, it represents 25% of pesticides and 10% of insecticides bought in the world.

8.5   Water consumption

Estimated total water used in wet processing of cellulosic fibres is 2.96 trillion litres considering water consumption of 100 l/kg of material. If we can reduce water consumption by ¾ (i.e., 25 l/kg), the saved water can provide drinking water for 2.34 billion people (assuming consumption of 2.6 litres per capita per day).

Water can be saved in dyeing in the following ways:

1. Reuse dye house water
2. Reduce reprocessing
3. Optimise rinsing and soaping processes
4. Reduce liquor ratio.

8.6   Polluting wet processes

The important environmental concerns related to textile wet processing are:

1. Chemical intensive wet processing—scouring, bleaching, mercerising, dyeing, printing, etc.
2. Use of heavy metals—iron, copper, lead, etc., found in dyestuffs auxiliaries, binders, etc.
3. Residual dyestuffs due to poor fixation of dyes and chemicals in effluent water.
4. Poly vinyl chloride (PVC) and phthalates used in plastisol printing paste.
5. Formaldehyde found in dispersing agents, resins, printing paste and colorant fixatives.
6. Dye effluent-wastewater issue.

Many of the chemicals used in textile processing can be recovered from waste water by membrane technology. The most problematic pollutant is the dye itself. Inherent to their purpose, dye molecules are designed to be resistant to degradation by light, water and many chemicals. Dye molecules can be decomposed in water by a range of chemical, physical and biological treatments. The most widely used technique is the oxidative process, where hydrogen peroxide is added to the water and activated by ultra violet light to oxidise the dye molecules. However, toxic sludge is produced, which has to be disposed of or incinerated.

8.6.1   Absorbable organic halogens (AOX)

It is a measured value for organically bound chlorine, bromine and iodine in a given substance. The AOX consent limit is likely to be as low as 2 ppm from a German drinking water directive (DIN 38409414, 1987) and as such compounds having high AOX values are to be used carefully, a few such products used in the textile industry are:

1. Chlorine-containing bleaching agent.
2. Shrink-proofing of wool with chlorine, the promising alternate being permonosulphuric acid.
3. Insect-proofing agent for wool.
4. Some types of carriers used in dyeing with disperse dyes.
5. Certain chromophores.
6. Some classes of reactive dyes.

8.7   Harmful chemicals

Some of the toxic and harmful substances used in textile industries and their eco-friendly substitutes are listed in Table. 8.1.

Table 8.1. Few harmful textile chemicals and their eco-friendly substitutes.

Existing chemicals	Uses	Proposed substitutes
Polyvinyl alcohol (PVA)	Yarn size	Potato starch or carboxy methyl cellulose (CMC)
Pentachlorophenol, formaldehyde	Size preservative	Sodium silicofluride
Carbon tetrachloride (CTC)	Stain removers	Detergent stain-removers Detergent (non-ionic, ethoxylates) and water miscible solvent (glycol ethers) mixtures Enzymatic stain-removers.

Calcium and sodium hypochlorite	Bleaching	Hydrogen peroxide, ozone at cold
Sodium silicate, phosphorous-based compounds	Peroxide stabiliser	Nitrogenous stabilisers
Nonyl phenyl ethylene oxide adducts (APEO)	Detergent emulsifier	Fatty alcohol ethylene oxide adducts, alkyl poly glycosides
Synthetic non-biodegradable surfactants	Various purposes	Sustainable and highly biodegradable surfactants from dextrins
Synthetic non-biodegradable surfactants + solvent	Coatings and degreasing	"sSolvo surfactants" acting both solvent and surfactant, derived from glycerol (biodiesel)
Dichloro and trichloro benzene	Carriers in dyeing	Butyl benzoate, benzoic acid
Kerosene	Pigment printing	Water-based thickeners
Formaldehyde	Finishing, dye fixing	Polycarboxylic acid, non-formaldehyde products
Sodium dichromate	Oxidation in dyeing	Hydrogen peroxide
Silicones and amino-silicones + APEO emulsifier	Softener	Eco-friendly softeners, wax emulsions
Functional synthetic finish	Finishing	Bees wax, aloe vera and Vitamin A (Hazardous Substance Research Centres/ South & South-west Outreach Program, "Environmental Hazards of the Textile Industry," *Environmental Update* #24, *Business Week*, June 5, 2005; http://www.oecotextiles.com/)

8.8   Green chemistry in textile industry

In paper and textile industries, efforts are being made to develop new greener methods, which result in reduction in energy, water usage, time in textile processing. Some examples of green approaches in various textile related industries are as follows:

The conventional chemical manufacturing processes are unsustainable because

1. Mostly carbon-based products are derived from fossil fuels, petroleum and coal which have limited supply.
2. Large amounts of waste increasing burden on the environment.

Environmental chemistry studies the effect of environmental pollutants, whereas green chemistry deals with new sciences and technologies to prevent the formation of any waste. In developing countries, although there is growing awareness about the ill effects of pollution, promotion of continual introduction of environmentally friendly products a methodologies in the chemical industry needs to be developed further. Usage of non-conventional technologies is highly popular in India. First in this list is the usage of microwaves. Further, the microwave chemists are turning their attention toward microwave-assisted dry-media reactions in order to minimize solvent usage, an added advantage to already established microwave chemistry. In addition to microwave-assisted reactions, ultrasonic and photochemical reactions are also used as non-conventional reaction technology. The strict application of the Euro norms and the drive for switchover to Compressed Natural Gas (CNG) as an alternative fuel are some of the examples of the endeavours made by India towards creating a greater eco-friendly environment.

8.9   Recycled textiles

Because textiles are nearly 100% recyclable, nothing in textile and apparel industry should be wasted. The textile recycling industry is one of the oldest and most established recycling industries in the world. Textile recycling materials may be pre-consumer or post-consumer (i.e., used garments or articles). The sorting categories of textile recycling by volume is represented by a pyramid structure, the base of which consists of used cloth market (48%), followed by conversion to value added new materials (29%), cut into wiping and polishing cloths (17%), landfill and incineration for energy (<7%). The peak of the pyramid is represented by "Diamonds" (1–2%) which have high value for antique quality or for other reason. Polyester fibre is one of the most non-biodegradable polymers which create environmental problems. Major revolution happened in 1993 when Wellman Inc. introduced the first polyester textile fibre made from post-consumer PET packaging: Fortrel® EcoSpun®. There are two broad types of recycled polyester namely:

1. Simply melted and re-extruded into fibres and
2. A multi-stage de-polymerisation and re-polymerisation to produce better quality yarn.

However, re-cycled polyester yarn is not always as good as virgin polyester. Colour consistency is difficult to achieve, particularly on pale shades. If the carpet fibres are made of polypropylene and they're held together with a polypropylene Licocene back-coating, the product can be reused simply by melting.

8.9.1   Greener dye and auxiliaries

The greener approaches are as follows:

1. Elimination of harmful azo dyestuffs.
2. Alternative synthesis for eco-friendly products.
3. Search for sustainable source such as natural dyes. They, in general, have poor to moderate light fastness. It was found that the natural additives Vitamin C (ascorbic acid) and gallic acid (found in stomach, tea leaves, oak bark and many other plants) were most effective in reducing the rate of fading in madder, weld and wood dyed cotton.

8.9.2   Biodegradable surfactants

By reacting dextrins with fatty acids and their derivatives, new sustainable and biodegradable surfactants have been formed. They have highly desirable physical properties including low foaming, good wetting and whitening ability, as well as excellent biodegradability. Queste reported that the researchers in France and Germany have jointly developed a new class of so-called "solvo surfactants" (which exhibit the properties of both solvent and surfactant and are commonly used in applications such as coatings and degreasing, as well as perfumery and inks) that are derived from glycerol, a renewable material from biodiesel.

8.10   Greener preparation and dyeing

Bio-processing can simply be defined as the application of living organisms and their components to industrial products and processes, which are mainly based on enzymes.

The application of enzymes in various stages of textile processing may be listed as follows:

- Desizing: amylase, lipase.
- Scouring: pectinase, cellulase.
- Bleaching: oxidoreductase. xylanase.
- Dyeing: oxidoreductase.
- Finishing: cellulase, oxidoreductase, lipase.
- Composting (biodegradation of textile wastes): cellulase, protease, nylonase, polyesterase.
- Delignification, decolourisation of dyes: laccases.

Some greener preparatory processes are as follows:

1. Purification of cellulose by extraction by carbon dioxide and ionic liquids;

2. High temperature water extraction of lignin;
3. Substitution of chlorine bleaching with non-polluting oxidants;
4. Carbon dioxide-based dry cleaning;
5. Elimination of ozone-depleting chemicals such as carbon tetrachloride (stain remover).

8.10.1   Mechanisms of enzyme

Enzymes are generally globular proteins and range from 62 amino acid residues in size. Most of the enzymes are much larger than the substrates, they act on and only a small portion of the enzyme (around 3–4 amino acids) is directly involved in catalysis (Carlier 2001). The active site continues to change until the substrate is completely bound, at which point the final shape and charge is determined (Etters and Anis 1998). The mechanism of enzyme catalysis is shown in Figure 8.2.

$$E + S \rightleftharpoons ES \rightleftharpoons EP \rightleftharpoons E + P$$

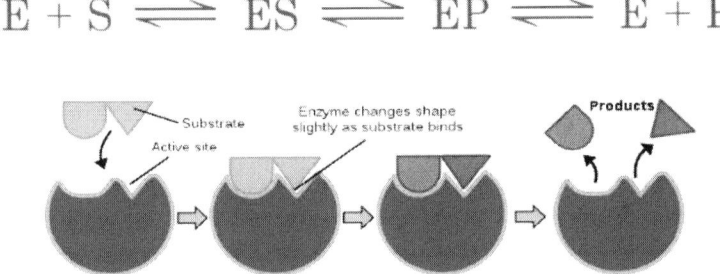

Figure 8.2. Mechanism of enzyme catalysis

The region that contains these catalytic residues, binds the substrate and then carries out the reaction is known as the "active_site". Enzymes can also contain sites that bind cofactors, which are needed for catalysis. Some enzymes also have binding sites for small molecules, which are often direct or indirect products or substrates of the reaction catalysed (Alat 2001). Enzymes can catalyse up to several million reactions per second. Enzyme rates depend on solution conditions and substrate concentration (Jayam et al., 2005). This binding can serve to increase or decrease the enzyme's activity, providing a means for feedback regulation. Textile processing has benefited greatly in both environment and product quality aspects through the use of enzymes. New enzymatic processes are being developed (cellulase, hemicellulase, pectinase, protease, xylanase, cutinase and lipase), which offer the potential to replace the use of other chemicals in textile preparation processes for natural and synthetic materials (Warke and Chandratre, 2003).

8.10.2   Enzymes in textile wet processing

The use of enzymes in textile processing and after care is already the best established example of the application of biotechnology to textiles in the near-to medium-term future. The applications of hydrolase and oxidoreductase enzymes used in the textile wet processing are given in Table 8.2.

Table 8.2. Application of hydrolase enzyme in fabric preparation.

Textile application	Enzyme name	Substrate
Starch desizing	Amylase	Starch
Stone wash/bio-polishing Bio- finishing for hand modifications Carbonisation of wood	Cellulase	Cellulose
Bio-scour replacing caustic soda	Pectinase	Pectin
In situ peroxide decomposition without any rinse in bleach bath	Catalase	Peroxides
Degumming of silk Bio-antifelting of wool/cotton wax	Protease	Protein molecules/ peptide bonds
Hydrophilicity of cotton and polyester	Lipases	Fats and oils
Discoloration of coloured effluent Bio-bleaching of lignin containing fibres like kenaf and jute. Bio-bleaching of indigo in denim for various effects	Laccase	Colour Chromophore and pigments
Bio-bleaching of wood pulp	Peroxidases	Colour Chromophore and pigments
In situ generation of hydrogen peroxide for bleaching of cotton	Glucose oxidases	Pigments
Discolouration of azo dyes effluent	azo reductase	Colour Chromophore and pigment
Discolouration of remould of basic dye effluent	Peroxidase ostreatus	Colour Chromophore and pigment

8.11   Desizing of cotton fabric

Sizing is a process used for the application of a film forming polymer to provide temporary protection to the warp yarns from abrasive and other types of stresses generated on the weaving machines in order to reduce the warp breakages. Sizing helps in forming a coating which encapsulates the yarn, embeds the protruding fibres and also causes some inter-fibre binding by penetration. The spun yarns being hairy usually require size add on exceeding 8–10%; depending on the fabric to be woven. Amylase assisted desizing of textile materials is carried out in machines such as jigger, jets, pad-batch and pad-stream ranges, employing different levels of mechanical agitations. Among various steps involved in enzyme reaction, hydrolysis of starch needs longer time, depending upon the activity levels of enzyme and temperature conditions used in desizing. The small addition of synthetic binder to starch causes plasticisation and increases the adhesion. Major drawback of the starch, brittleness of the film, can be reduced by providing internal plasticisation (Shah et al. 1976; Moghe and Khera 2005).

For cotton fabrics, traditional desizing is being carried out by high temperature washing process and high concentrations of surfactants. As the process proceeds, the viscosity of the washing liquor rises rapidly because sizing agents dissolve. Therefore, large amounts of hot water are required. The use of industrial enzymes for desizing represents a major improvement because they cleave the biopolymers used a sizing agents into sub units as oligosaccharides. In some cases, this is a first step in bio-preparation of cotton fabrics (Buschie-Diller et al. 1994; Tzanko et al. 2001). The mechanical properties of synthetic size materials are better than starch based materials. Among different synthetic sizes, PVA exhibits overall better performance (Bayard 1983). Common waxes do not inactivate amylases but prevent quick wetting, penetration of enzymes and, other factors that affect the efficiency of size removal include viscosity of starch, amount of size applied, fabric construction and method of washing-off (Shamey and Hussein 2005).

Anaerobial microbial cultures in the desizing of cotton fabrics was studied and reported that the bioreactor from wastewater and microbial culture for desizing of cotton was performed upto 73% desizing efficiency at 55°C and time of 60 min and it depends mainly on the temperature and reaction time of the process and need for optimising the process conditions for achieving better desizing and low energy (Heikinheimo et al. 2003). Several factors affecting the starch-size removal were studied and are (a) the effectiveness of enzymatic desizing can be enhanced by raising the desizing temperature up to 70°C; prolonging the desizing time up to 60 min; increasing the material-to-liquor ratio up to 20:1; increasing the Aquazym® 240-L dosage up to 6 g/l; treating at pH 7 (Nabil et al. 2004; Csiszar et al. 2001). Most enzymes are sensitive to pH and have specific ranges of activity. The pH can stop enzyme activity by denaturating (altering) the three dimensional shape of the enzyme. Most enzymes function between a pH of 6 and 10 (Li and Hardin, 1999). All enzymes work within a range of temperature specific to the organism. Increases in temperature generally lead to increase in reaction rates (Daniel et al. 2010). Composition, properties of starch (Azevedo et al. 2003; Aranjo et al. 2004; Moghe and Khera 2005), ingredients added in size mix (Tomazic and Klibanov 1969; Shah and Sadhu 1976; Lange 1997; Azevedo et al. 2003; Declerek et al. 2003) and process conditions employed in desizing (Fetouh et al. 1974; Khalil et al. 1974; Levene and Prozan 1992; Mori et al. 1997; Ibrahim et al. 2004) have marked influence on the efficiency of desizing.

8.12   Scouring of cotton fabrics

The scouring process is carried out to make the material hydrophilic, before it undergoes other processes like bleaching, dyeing and printing in the textile wet processing (Holme 2001). Enzymes are substrate specific biocatalysts; they operate best at ambient pressures, mild temperatures and often at a neutral pH range. Enzymes are gaining an increasingly important role as a tool in various wet textile pre-treatment and finishing processes (Alat 2001; Anon 2001; Carlier 2001). The concentration of alkali used and the time and temperature conditions needed depend on the desired quality of the scoured fabric. Other chemicals for instance, wetting agents, emulsifying agents and chelating agents (Nallankilli et al. 2008; Tyndall 1996) are included in typical preparation baths for scouring. Wetting agents act by reducing the surface tension of water enabling improved penetration of the chemicals into the cotton fabric. Emulsifying agents assist in removing waxy materials. Chelating agents remove polyvalent metal ions such as calcium, magnesium, iron or other salts that can have a harmful effect on subsequent wet-processing operations. Effective scouring is essential for subsequent processing of any cotton made substrate, regardless of its natural source. Even today, alkaline scouring of cotton is still the most widespread commercial technique for removing or rupturing the fibre cuticle to make the fibre absorbent for the cotton processing. Although sodium hydroxide is used generally for the scouring, sodium carbonate and calcium hydroxide are mentioned as a scouring agent (Hsieh and Cram 1999). Scouring of cotton fabric is typically done with a hot solution (90°C–100°C) of sodium hydroxide (± 1 mol/l) for up to 1 h (Emilla Csiszar et al. 1998).

A desired hydrophilicity during the scouring can be achieved by removing non-cellulosic material from the cotton fabric, especially from the cuticle (waxes and fats) and the primary wall (e.g., pectin, protein and organic acids). More precisely scouring not only removes non-cellulosic material from cotton fibres but also removes substances that have adhered to the fibres during the production of the yarn or fabric. Substances like, dirt, lint, pesticides, oils, and any sizing agent applied to yarns to facilitate weaving (Eisisi et al. 1990; Ammayappan et al. 2003). The scouring process requires large quantities of chemicals, energy and water and is rather time consuming (Yonghua and Hardin 1997). Biocatalysts have proven to be a flexible and reliable tool in wet textile processing and a promising technology to fulfil the expected future requirements. Enzymatic scouring has been investigated extensively by various institutes and laboratories now for nearly one decade (Hartzell and Hsieh 1998; Emilla Csiszar et al. 1998; Qiang Wang et al. 2006). Owing to the high sodium hydroxide concentration and its corrosive nature, intensive rinsing is required that leads to high water consumption. The use of high concentrations of sodium hydroxide also requires the neutralization of wastewater, which requires additional acid chemicals. Furthermore, the alkaline effluent requires special handling because of very high Biochemical Oxygen Demand (BOD, also called Biological Oxygen Demand) and Chemical Oxygen Demand (COD) values. Apart from the above wet processing problems, the biggest drawback of alkaline scouring is a non-specific degradation of cellulose that produces fabrics of lower tensile strength and therefore of lower quality (Wang et al. 2006).

Enzymes are substrate specific biocatalysts; they operate best at ambient pressures, mild temperatures and often at a neutral pH range. Enzymes are gaining an increasingly important role as a tool in various wet textile pre-treatment and finishing processes (Alat 2001; Anon 2001; Carlier 2001). Pectinases are either endo-acting, cutting the polymer at random sites within the chain to give a mixture of oligomers or exo-acting attacking from one end of the polymer and producing monomers or dimers, identified by the rate of release of reducing sugars (Friend and Chang 1982). In enzymatic scouring of cotton, non-ionic surfactants are used to overcome hydrophobicity of the substrate, which assist enzymes to penetrate through micro-pores or cracks and help them to orient themselves in favourable positions for catalytic actions (Sahin and Gursoy 2005), while ionic surfactants complex with enzymes and disrupt their structure to different extents (Tzanov et al. 2000). Pectinase is an enzyme that catalyses hydrolysis/depolymerisation of the glycosidic bonds in the pectin polymers, classified according to their preferential substrates (high or low methyl esterified pectin and polygalacturonic acid/pectate) and their reaction mechanism (Kristensen 2001; Jayam et al. 2005). Cotton fabrics, using cellulases, is aimed to remove cellulosic impurities, individual and loose fibre ends that protrude from fabric surfaces and to provide an enhanced appearance and handle, with or without the aid of mechanical agitations but without degrading the properties of the fabrics significantly. Cellulase enzymes are complex mixtures of three major constituents enzymes namely, endo l-4 α D glucanases (ED) (EC 3.2.1.4), which randomly cleave internal glucosidic bonds, 1-4, α D glucan cellobiohydrolases (CBH) (EC 3.2.1.91), which cleave them into cellobiases. Hydrolysis of cellobiose into the glucose end product is completed by β glucosidases or cellobiases (EC 3.2.1.21), which split cellobiose units into soluble glucose monomers and complete hydrolysis of native celluloses, largely, depends on the synergistic actions of these three component enzymes.

8.13   Enzymatic bleaching

Many attempts have been made to utilise various enzymes that belong to oxidoreductases in bleaching of cotton fabrics (Nalankilli and Sundar 2002; Tzanko et al. 2002; Opwis et al. 2006; Diller and Traore 1998; Anis et al. 2008; Anis et al. 2009) and post-bleaching processes (Jensen 1998). Conventional bleaching methods have been reviewed in many occasions, (Shenai 1996; Dickinson 1979; Maekawa et al. 2007), oxidative chemical pre-treatments are effective in degrading colourants and other impurities, though such methods often lead to oxidative degradation of substrates. However, peroxide bleaches are often referred to as the "colour safe bleaches", due to minimal degradation of substrates. Peroxidases are used to activate oxidizing agents like hydrogen peroxide, however, rapid deactivation of these agents in bleaching process does not guarantee satisfactory bleaching effects (Bernards et al. 2004). Mechanical actions, winch machines and jet systems, between fabrics and equipment or surface to surface contact of fabrics enhance reactivity of cellulases by improving two way mass transfers and enhance weight loss, removal of weakened fibres from surfaces of yarns and fabrics, thereby facilitating a clean surface to the fabrics (Ogiwara and Arai 1968; Tyndall 1996; Paulo and Almeida 1994; Lee et al. 1996; Gama et al. 1998; Andreaus et al. 1999; Traore and Diller 1999; Lee et al. 2000; Heikinheimo et al. 2003; Ramkumar and Abdalah 2001). Material to liquor ratio of process bath alters the efficiency of all the components exhibited by weight loss values and, little changes are observed in the range of 1:10–1:40 (Paulo et al. 1996). Besides the nature of substrates, efficiency of hydrolysis is also influenced by process conditions (Hemmpel 1991; Tyndall 1996; Paulo et al. 1996; Andreaus et al. 1999), co-reactants present in the process. Higher weight loss values (7.0–15.3%), better drop absorbency (~1 s), dye absorption (*KIS* values of 7.45 against 6.9 of acid desized samples) have been reported with higher concentration of enzymes and longer incubation time (Dalvi et al. 2007). Whenever a very high whiteness in the fabrics is not required, desizing is combined with bleaching and scouring, incorporating protease, cellulase and pectinase enzymes (Lange et al. 2001; Miller et al. 2003).

Integrated desizing and scouring using α-amylase and polygalacturonate lyase process involves two steps, in which fabrics are desized first, using α-amylases, then by a combination treatment of amylase and pectate lyase at 45–55°C at pH of 8.5–9.0, followed by washing in presence of chelating agent at 90–100°C (Lenting 2008). Simultaneous desizing and scouring using amylase and pectinase obtained from single source, i.e., *Bacillus* and different sources have been attempted in the past (Lenting and Warmoeskerken 2004; Dalvi et al. 2007). Combinations of pectinases with protease, hemicellulase, cellulase and lipase have been attempted, which are not efficient in scouring process when used alone. Combinations of amylases with other enzymes, preparatory chemicals have been attempted in the past to combine scouring or bleaching (Etters 1999; Tzanov et al. 2000; Lu 2005; Opwis et al. 2006; Kuilderd and Wu 2008; Lenting 2008). In the case of amylase desizing, addition of hydrogen peroxide improves whiteness, while neutral cellulases increase weight loss and desizing efficiency. Alkaliphile amylases with sodium hydroxide and hydrogen peroxide have been recommended for a combined desizing–scouring–bleaching process (Etters 1999; Csiszar et al. 2007). Ultrasonic technique holds a promise in applications in the field of textiles. Ultrasonic represents a special branch of general acoustics, the science of mechanical oscillations of solids, liquids and gaseous media. Ultrasound can enhance a wide variety of chemical and physical processes, mainly by generating cavitation in liquid medium. The sonicator used is of 20 kHz frequency which is found to be suitable for inducing cavitation (Abramov 1998; Kamel et al. 2005).

Some greener dyeing processes are listed below:

a) Improvement in the existing dyeing processes
1. Optimise processes (to reduce time and energy consumption).
2. Reduce consumption of water, electrical power, steam consumption.
3. Optimise dye/chemical costs.
4. Eliminate reprocessing and shade correction.
5. Sulphur dyeing: substitution of hazardous sodium sulphide with sustainable, nontoxic, biodegradable, cost-effective reducing sugars.
6. Reactive dyeing: treatment of cellulose with cationic, nucleophilic polymers enables dyeing at neutral pH without electrolyte addition—3-chloro-2-hydroxypropyl-trimethylammonium chloride (CHTAC), copolymer of diallyldimethylammonium chloride and 3-aminoprop-1-ene and copolymer of 4-vinylpyridine quaternised with 1-amino-2-chloroethane.
7. In chemical-free denim processing, laser technology is used to burn away the surface of the dyed denim fabric or a pair of jeans on a mannequin to replicate an authentic worn look. The laser system is very quick and a pair of jeans can take as little as 15 s to process.
8. Right–First–Time (RFT) dyeing: It is also termed as "no addition" dyeing or "blind dyeing". Elimination of the inspection stage made a significant saving. Twenty factors which must be monitored or controlled to achieve RFT processing in the dyeing process have been identified.

b) New dyeing technologies with minimum environmental impact
1. About 90% dye fixation on batch wise cellulose dyeing with polyfunctional dyes.

2. Cold pad-batch dyeing, rapid dyeing techniques and better machine design.
3. Economic continuous dyeing methods.
4. Supercritical carbon dioxide dyeing.

8.14   Formaldehyde-free finishing agents

The most widely used crosslinking agents in DP finishes, *N*-methylol agents or *N*-methylolamides fall in the category of formaldehyde reactants. The release of formaldehyde vapours is a problem with those agents. It depends on the reactant types, the catalyst types, the condition of the treated fabrics, and the additives in the impregnating bath and most importantly the time and temperature of cure.

The Occupational Safety and Health Administration (OSHA) have set the upper limit for formaldehyde in air at 0.75 parts per million averaged over an 8-hour work shift. Formaldehyde is a carcinogen to animals. Some formaldehyde-free DP finishes are as follows:

1. Cyclic addition of glyoxal with NN/dimethyl urea, namely DHDMI (1,3 dimethyl-4,5-dihydroxyethyleneurea)
2. Polycarboxylic acids (PCA)—their main drawback is loss of tensile strength due to acid-catalysed cellulose chain cleavage. The most important PCA reactants are butanetetracarboxylic acid (BTCA) and citric acid (CA). BTCA, in the presence of sodium hypophosphite, provides the same level of durable press performance as conventional DMDHEU reactant, but it is quite costly.

8.15   Eco-friendly flame retardants

An interesting development in the field of flame retardancy is the use of polymer nano-composites as a substitute of toxic brominated flame retardants (BFR). Nano-composites may be described as two-phase materials, consisting of a dispersion of appropriate filler (on a nanometre scale) through a polymer matrix. In the case of polymer-layered silicate (clay) nano-composites, only a very small amount of filler (2–10 weight-%) is required for the material to be flame-retardant.

8.16   Conclusion

Globalisation is inevitable and unavoidable under the present world economic situation. Many industries are affected positively or negatively with the globalisation trend. Sustainability has become an essential attribute of today's textile and clothing industry, the process of transforming textile industry into more sustainable one is very sensitive, needs a lot of knowledge, skills and commitment. Green chemistry provides a technical solution to many environmental problems. It is effective due to design stage efforts, starting at the molecular level lets one to design out the hazardous properties and to design in environmentally appropriate features. Global awareness of the real price of clothing is growing and there are increasing numbers of cases of people experiencing health problems such as rashes, allergies, respiratory and concentration problems due to chemical sensitivities. There has been a growing concern over apparel brands in improving their environmental impact and the economic responsibility throughout their supply chains.

8.17   References

1. Abada, E.A.E., (2008), Production and characterization of a mesophilic lipase isolated from Bacillus stearothermophilus AB-l, Pakistan Journal of Biological Sciences, 11(8): 1100–1106.
2. Abramov, O.V., (1998), High intensity ultrasound: theory and industrial applications, Gordon and Breach Publications, London. ISBN 9789056990411, 58–94
3. Abu, E.A., Ado, S.A. and James, D.B., (2005), Raw starch degrading amylase production by mixed of aspergillus niger and saccharomyces cerevisae GRMVTH on Sorghum Pomace, African Journal of Biotechnology, 4(8): 785–790.
4. Agarwal, P.B., Nierstrasz, V.A., Santuer, B.G.K., Gubitz, G.M., Lenting, H.B. and Varmoeskerken, M.M. (2007), Wax removal for accelerated cotton scouring with alkaline pectinases, Biotechnology Journal, 2(3): 306–315.
5. Aiteromem, A.L., (2008), Sustainable textile manufacturing with new enzyme based processes, International Dyer, 7: 15–19.
6. Ajayl, A. and Fagade, B., (2008), Growth pattern and structural nature of amylases produced by some Bacillus species in starch substrates, African Journal of Biotechnology, 5(5): 440–444.
7. Alat, D.V., (2001), Recent developments in the processing of textiles using enzymes, Colourage, 48(2): 33–36.
8. Ali, S.L. and Khan, A.F., (2005), Development of stabilized vegetable amylases for enzymatic desizing of woven fabric with starch containing sizes, Pakistan Journal of Scientific and Industrial Research, 48(2): 128–130.
9. Allen, A.M., Foulk, J.A. and Gamble, G.R., (2006), Fourier transform infrared spectroscopy analysis of modified cotton trash extraction, Proceedings of Beltwide Cotton Conferences, Texas, 1–10.
10. Almeida, C., Branyik, T., Ferreira, P.M. and Teixeira, J., (2003), Continuous production of pectinase by immbolized yeast cells on spent grains, Journal of Bioscience and Bioengineering, 96(6): 513–518.

11. Alves, M.H., Takaki, G.M.C., Porto, A.L.F. and Milanez, A.I., (2002), Screening of Mucor spp for the production of amylase, lipase, polygalacturonase and protease, Brazilian Journal of Microbiology, 33(4): 325–330.

12. Ammayappan, L., Muthukrishnan, G. and Saravana, P.C., (2003), A single stage preparatory process for woven cotton fabric and its optimization, Man Made Textiles in India, 47(1): 29–35.

13. Andersen, L.N., Schulein, M. and Lange, N.E.K., (2001), Pectate lyase, US Patent No. 628 4524.

14. Andersen, L.N., Schulein, M., Lange, N.E.K., Ranvan, M.E.B., Fler, S.M., Glad, S.O., Kauppinen, M.S., Schnorr, K. and Kongsbak, L., (2002), Pectate lyases, US Patent No.6 368843.

15. Andrade, V.S., Sarubho, L.A., Fukushima, K., Myaji, M., Nishimura, K. and Takaki, G.M.C., (2002), Production of extracellular proteases by Mucor circinelloides using D-glucose as carbon source substrate, Brazilian Journal of Microbiology, 33(2): 106–110.

16. Andreaus, J., Azevedo, H. and Paulo, A.C., (1999), Effects of temperature on cellulose binding activity of cellulase enzymes, Journal of Molecular Catalysis B: Enzymes, 7(4): 233–239.

17. Anis, P., Davulcu, A. and Erien, H.A., (2009), Enzymatic pretreatment of cotton. Part 2 - Peroxide generation in desizing liquor and bleaching, Fibres and Textiles in Eastern Europe, 17(2): 87–90.

18. Anon, (2001), Cotton preparation - A new enzymatic concept, Novozyme Catalogue, 23105-0 l.

19. Anon, (1982), Evaluating combined preparation processes for energy and material conservation, Textile Chemists and Colorists, 14(1): 23–35.

20. Anon, (1954), Raw cotton, Textile Research Journal, 24(8): 744–747.

21. Anto, H., Trivedi, U. and Patel, K., (2006), Alpha amylase production by Bacillus cerens MTCC 1305 using solid state fermentation, Food Technology Biotechnology, 44(2): 241–245.

22. Appleyard, H. (1953), Enzyme actions in desizing, The Dyer and Textile Printer, 5(2): 685–688.

23. Aranjo, M.A., Cunha, A.M. and Mota, M., (2004), Enzymatic degradation of starch based thermoplastic compounds used in Protheses: Identification of the degradation products in solution, Biomaterials, 25(13): 2687–2693.

24. Arguelles, M.E.A., Rojas, M.G., Gonzauz, G.V. and Torres, E.F., (1995), Production and properties of three pectinolytic activities produced by Aspergillus niger in submerged and solid state fermentation, Applied Microbiology and Bioteclmology, 43(5): 808–814.

25. Asgar, M., Asad, M.O.J., Rahman, S.U. and Legge, R.L., (2007), A thermostable alpha amylase from a moderately thermophilic Bacillus subtilis strain for starch processing, Journal of Food Engineering, 79(3): 950–955.

26. Anastas, P. and Warner, J.C., (1998), Green Chemistry, Theory and Practice, Oxford University Press, New York. ISBN-13: 978-0198506980, 48–72

9

Reduction of carbon footprints in apparel industry

Dr. M. Parthiban and Dr. P. Kandhavadivu

Department of Fashion Technology, PSG College of Technology,
Coimbatore-641 004
Email:parthi111180@gmail.com

Abstract: "Global warming" and "carbon footprint" are a buzz word now. Its importance and the consequential long-term devastating effects of "climate change" on the environment, habitat and even the existence of our mother Earth are widely discussed. This warming of atmospheric temperature is attributed to the emission of Greenhouse Gases (GHG)—Carbon dioxide (CO_2), methane, nitrous oxide and fluorocarbons are the major contributors. The recent years have witnessed exponential increase in the emission of Greenhouse Gases (GHG) raising the atmospheric temperature. It is reported that there is about 6% rise only in the year 2010 (releasing about 500 mega-newton/megatonne) majority of which is attributed to the top three pollutants of the world—China, the USA and India. The GHG emission is caused by the production and consumption of fuels, manufactured goods, materials, wood, roads, and services. For simplicity of reporting, it is often expressed in terms of the amount of CO_2, or its equivalent of other GHGs, emitted. Just as walking on the sand leaves a footprint, burning fuel leaves CO_2 in the air, which is called a "Carbon Footprint". Hence, the carbon footprint basically relates to the amount of carbon released into the air based on the fuel consumption.

Keywords: carbon footprint, methods, machinery, process & concepts

9.1   Introduction

- Carbon footprint—the total set of greenhouse gas emissions caused by an individual, event, organisation or product, expressed as carbon equivalents.
- Carbon footprint is one of the family of footprints which includes water footprints and land footprints.
- The concept name originated from ecological footprint, which was developed by Rees and Wackernagel in 1990s.

- An individual's, nation's or organisation's carbon footprint can be measured by undertaking a GHG emissions assessment or other calculative activities denoted as carbon accounting.
- Carbon accounting can be done either from the carbon emissions or the indirect carbon emissions.
- Direct carbon emissions denote the GHG emitted on using the various forms of energy like nuclear, hydro, coal, gas, solar cell and wind generation technology among which hydro, wind and nuclear power produce the least CO_2 per kilowatt-hour.
- The indirect carbon emissions include the calculation of GHG emissions involved in each step of the production of any product.
- Carbon footprint is the sum of all emissions of CO_2 which were induced by the activities in a given time frame.
- The usual time period taken is a year.
- The other greenhouse gases like methane and ozone are also taken into account for the carbon footprint. They are converted into the amount of CO_2 that would cause the same effects on global warming (called equivalent CO_2 amount).[1]
- Sometimes carbon footprints can also be expressed in kilogram carbon. Kilogram CO_2 can be converted to kilogram carbon by multiplying the final value by a factor 0.27.
- There are carbon footprint calculators available online which allows to store individual activities and see the amount of CO_2 created for each individual activity.

9.2   Role of carbon footprint in apparel industry

- Reveals how much CO_2 in total is emitted along the value of chain of a product.
- Total set of greenhouse gas emission caused by an organisation or product.
- It is calculated for the time period of a year and expressed in terms of the amount of CoO_2, or its equivalent of other GHGs emitted.
- As greenhouse gases produced by human activities accumulate and their concentration increases in the atmosphere, it causes global warming.[2]

9.3   Factors behind textile CO_2 emissions in textile industry

- The majority of fibres produced are synthetic such as petrochemical-based nylon and polyester, and chemical-treated rayon, use massive amounts of energy to create.

- Conventional cotton is also heavily detrimental to the environment. Cotton growth requires intensive use of pesticides, chemical fertilizers, and water.
- The dyeing and bleaching of fabrics involves chemicals, energy, and huge amounts of water. Approximately 1 million tons of chemical dyes are used every year.
- The wet finishing process uses huge amounts of water and energy.
- Wet treatment of textiles like desizing, pre-washing, mercerising, dyeing, printing, etc., includes a lot of chemical applications on the fibres or fabric. Water is used at every stage in fabric manufacturing.[3]
- Some fibres need to be bleached with chlorine before dying. This causes organo-chlorine compounds to be released, which are very dangerous to the environment. It takes between 10% and 100% of the weight of the fabric in chemicals to produce that fabric.
- From dyes to transfer agents, around 2000 different varieties of chemicals are used in textile industries.
- Although, with 16% of the global population, India's share of CO_2 emissions is only 3.11%.

9.4   Methods to reduce carbon footprints in textile processing

Various ways and methods for reducing the carbon footprint during textile processing have been reported and widely published.[4] Commercially viable products are available in market and being supplied by many organisations. Some of the major areas of work are as follows:

A. Machinery/equipment related
- Use of low and ultra-low liquor ratio machines—to reduce consumption of water during pre-treatment—dyeing and post-dyeing wash-off sequence. Simultaneously reducing the energy required for water testing at various processing steps and effective load on the effluent treatment.
- Pre-heating of water by solar panels to reduce consumption of others non-renewable energy sources.
- Adequate insulation of dyeing, drying and stenter machines and appropriate heat recovery systems to avoid undesired energy loss.
- Recycle and reuse of process water and alkali by installing adequate filtration process.

B. Process related
- Combined scour and bleach process, combined peroxide neutralising and bio softening process, one bath one step dyeing of P/C blends,

etc., so as to reduce number of textile processing stages and thereby reduce consumption of water and energy.

- Cold pad batch preparation and dyeing for energy conservation.
- Continuous processing of knits.
- Pad/dry vs. pad/dry/steam, minimising steam and water consumption during washing processes and minimising number of during processes.
- Foam dyeing, finishing and coating.
- Improving Right-First-Time and Right-Every-Time dyeing performance.

C. Chemicals and dyes

- Use if enzymes—biodegradable and non-degradable corroding for desizing, scouring, bleach neutralising, bio-softening and post-dyeing wash-off. Suppliers and formulators of enzymes are offering specialised products for combined processes to reduce number of processing steps.
- Cationisation of cotton for salt-free dyeing with reactive and direct dyes.
- High fixation reactive dyeing which reduces salt for exhaustion.
- Digital inkjet printing.
- Low-temperature curing pigment printing.

D. Waste water treatment

- Use of physical, biological and activated carbon systems.
- Waste water treatment sludge used/sold for fuel.

Atul Ltd, pioneer in manufacturing of Dyestuffs in India and a major producer of Dyes, Pigments and Textile chemicals of International repute is a member of Ecological and Toxicological Association of Dyes and Organic Pigments Manufacturers (ETAD) and supplies products conforming to various global safety and Eco Conformance standards like GOTS, REACh, Blue Sign, etc. Atul has already initiated and developed products and processes to reduce carbon footprint not only during manufacturing of dyestuffs but also during the textile processing. Use of renewable energy source based on hydroelectric power of 45 MW, control of gaseous emissions by use of sophisticated containment devices and a modern Effluent Treatment Plant(ETP) and water treatment plant for recycling and reusing of water during dyestuff manufacturing. Atul has reduced greenhouse gas emission by approx. 150,000 MT/year through innovative technologies and received host country approval for three projects under Clean Development Mechanism (CDM).

Being the largest manufacturer of Vat, Sulphur and Reactive dyes in India, the focus is on cotton processing and products for reducing water and energy during colouration and subsequent processes.[5] Given below are few initiatives and achievements in this direction.

1) Tulacon C process: These are specialty formulated Vat dyes in liquid form, developed for application by a simple and efficient process of Pad–Dry–Cure on woven cotton fabric and its blends in open width form for a wide gamut of pastel shades. The advantage of this process over conventional Vat dyeing by exhaust or Pad-Dye Pad-Steam (PDPS) continuous method is mainly in terms of substantial water and energy saving. The conventional Vat dyeing on Jigger consumes about 15–20 l/kg (considering MLR of 3 and light shade) consisting minimum steps of—dyeing, rinsing, oxidation, soaping and neutralisation wash. This process requires temperature of about 50–60°C depending on class of Vat dyes for 1–2 h depending on the fabric length during dyeing and further during oxidation and soaping. Additionally, energy is consumed during fabric drying and thermo-chemical finishing. In case of conventional PDPS process, it involves dye padding–drying, chemical padding and steaming followed by a wash-off sequence on continuous washing range, drying and finishing. This too consumes substantial amount of water during washing-off and energy for intermittent drying, steaming and finishing process. In comparison to this, the specific advantage of Tulacon C range and process is in terms of

- Ready to use and easy to handle liquid form
- Simple application process
- No intermittent washing and no post-dyeing wash-off sequence
- A combined dyeing and finishing process. As the dye bath chemicals confer desired soft feel minimising post-dyeing thermo-chemical finishing step
- Considerable saving in water, time and energy
- Excellent lab to bulk reproducibility
- No wash-off—no effluent generation—environment friendly.
- Tulacon C—X g/l (up to 7 g/l)
- Tulachem ATB—10–20 g/l
- Tulachem ATS—5–10 g/l
- Glauber's salt—5–10 g/l (optional)

Adjust pH in between 5–6 with Tulachem Demin C (non-volatile, organic acid having buffering capacity). Pad (60–70% expression)—Dry at 110–130°C—Cure at 170°C for 30–45 s (cotton)/190°C for 30 s (polyester/cellulosic blends). Infra-red drying can be introduced prior to hot flue drying. A general and indicative carbon footprint in terms of water and energy saving based on estimations of usage during dyeing, post-dyeing wash-off and effluent treatment is considered to be:

Vat dyeing method: Water in ielitre energy effluent in litre
Exhaust—Jigger 15–20 5–6 KW/h 15–20
PDPS—continuous 10 2–3 Kw/h 10
Tulacon C—Almost nil 0.5–1 Kw/h almost nil.

1) Tularevs XL dyes—A high tinctorial, high fixative, low wash-off, long lasting, sustainable reactive dyeing system. Compared to the conventional reactive dyes, which exhibit comparatively low dye exhaustion and fixation levels and proportionately high wash-off of un-reacted hydrolysed dyestuffs, the molecular re-engineered Tularevs XL reactive dyes exhibit high colour yield and less wash-off. These warm dyeing dyes have similar dyeing profile which helps achieve uniform level dyeing and right-first-time (RFT) performance. Owing to the high fixation levels, this compact range of dyes covering wide shade gamut achieves outstanding wet fastness properties, long lasting colour shades. The low wash-off and less ensures low effluent generation. Though the exact impact and saving in carbon footprint is not yet ascertained, the overall water and energy saving due to short dyeing and wash-off cycle is practically proven. Thus the advantages envisaged are:

a) Increase productivity: Based on shorter process cycle, RFT (right-first-time) and RET (right-every-time) behaviour.

b) Cost optimisation: High-colour strength for optimum shade built-up, easy wash-off resulting in reduction in utility cost.

c) Eco-conformance: Meeting international product safety standards and eco-norms.

d) Optimum fastness: Satisfies stringent fastness and quality expectations of major brands.

Some other products and processes also provides considerable reduction in consumption of water and energy.

2) Rucoflow Cold Pad Batch (CPB): A ready to use, easy to handle liquid buffered alkali recommended for use in CPB Reactive dyeing is used for print fixation for partial or complete replacement of sodium silicate. Rucoflow CPB confers optimum alkalinity desired for dye fixation, assists ease of dye penetration inside the fibre, improves colour yield, ensures uniformity of dyeing and is easy to wash-off. Thus, helps in optimising the water and energy consuming wash-off sequence required during silicate fixation process. Additionally the free flowing liquid form ensures the quantity for the auto dosing systems. The specific advantages of reactive dyeing with Rucoflow CPB system over the conventional.

CPB dyeing system are:

- Sodium silicate being viscose fluid, available in varying Na_2O: SiO_2 ratios difficult to control desired alkalinity for reactive dye fixation.

- High viscosity and presence of impurities in commercial grades affect choking/clogging of dosing systems.
- Difficult to wash-off—requiring large quantity of water.
- Tends to impart undesired harsh handle, surface feel—requiring subsequent higher dosage of finishing softeners.
- Difficult effluent treatment, especially in case of intended water recycling by RO process—blocking of RO membranes.

3) Rucogen SOP—a novel washing-off agent specifically designed to minimise reactive dye soaping process and improve wet fastness properties. Generally, in conventional washing-off sequence, depending on the depth of shade about 4–8 post-dyeing wash cycles involving soaping and intermittent hot and cold washes are commercially practiced. Rucogen SOP (**S**ave **O**ur **P**lanet) is a unique Ter-polymer derivative designed to ensure optimum removal of unfixed, un-reacted or hydrolysed dye from the fibre without adversely affecting the adequately formed dye-fibre bond and avoids re-deposition or back staining. Thus, lowers the carbon footprint by reducing number of wash-off baths and load of coloured water effluent. Rucogen SOP includes the following merits:

- Has affinity for the dyestuff, being a mildly cationic polymeric compound
- High emulsifying and dispersing property helps prevent agglomeration of unfixed dye and wash-off residues in the bath
- Forms a stable metal ion complex and avoid adverse effect on dyestuff in case of presence of water hardening agents
- Exhibits dye transfer inhibition property by not allowing the dye re-deposition or back staining
- Improves wet fastness properties.

Many such innovative products and processes are being developed by researches and organisations across the globe for minimising the processing steps, offering alternative sustainable technologies for ultimate reduction, reuse and recycling of water and conservation of energy to reduce carbon footprint. The sustainability aspects, like water and energy saving are key concerns of the textile industry and novel technologies, in turn, show the ways to achieve such savings.

9.5   Conclusion

The global textile industry has taken several strides towards reducing its carbon footprint and meeting the challenges of building a more sustainable future. At the same time there is a growing awareness of environmental issues among consumers who are increasingly now increasingly insisting on textile products complying with environmental standards. These complementary trends will hopefully continue to drive the industry toward offering the consumer products that are not only red, blue, white, etc., but also green. Beyond fibre production, the dyeing and finishing sector is the largest energy and water consumer in the whole textile chain and has the highest potential for energy and water savings and efficiency improvements. Action is needed, but the industry cannot do it alone. National and multinational governments should support the industry with incentive plans to change old technology with modern equipment.

9.6   References

1. http://www.investopedia.com/terms/c/carbon_credit.asp#axzz1sqTy0KQF on Nov 26, 2015.
2. Emerging Issues in Apparel Trade, Sustainable Development and Carbon Neutrality Report, Apparel Export Promotion Council, Retrieved fromwww.aepcindia.in.
3. UK Launches First Carbon Footprint Label for Retail Clothing (Environmental Leader, March 27, 2009)
4. http://www.environmentalleader.com/2009/03/27/uk-launches-first-carbon-footprint-label-for-retail-clothing on Aug 24, 2016.
5. Market Emerging for Green Textile Chemicals." (Sustainable Plastics) http://www.sustainableplastics.org/news/market-emerging-green-textile-chemicals, Jan 1, 2018.

10
Eco-testing of apparel products

Amutha. K

Assistant Professor, Department of Textiles and Apparel Design, Bharathiar University, Coimbatore – 641 046

Email: amuthatad@buc.edu.in

Abstract: Textiles serve to be an indispensable part of human life. Textiles are used by human being to protect oneself from environmental elements such as heat, rain, wind, snow, etc. Hence the fundamental function of textile materials is protection. But the current situation is quite contradictory: these materials meant for protection pose numerous health hazards to human and affect the eco-system in various ways. Wide use of synthetic fibres, harmful dyes, mordants and other chemicals are the reasons for the health hazards and environmental pollution. Hence, the harmful substances need to be identified, quantified, permissible limits need to be declared, and finally tested and analysed to ensure safety of consumer.

Keywords: eco-parameters, banned amines, harmful chemicals, heavy metals, textile testing

10.1   Introduction

Consumers of this technological era are well aware of the impacts of the products they use in their daily life. It is not just the economic factor that decides the choice of purchase; the other criteria are health, environment and social concerns. Consumers' choice is shifting towards green or organic products. Hence it becomes necessary for the manufacturers, brands, and businesses to ensure user- and environment-friendliness of the products.

"Eco" is a prefix mostly relating to ecological or environmental terms. Hence eco-testing may be defined as the testing of the consumer safety and environment-friendliness of the product. Textiles play an important role in everyone's life. But the production and processing of textiles pose a great threat to the consumer and the environment due to the presence of various hazardous substances as a result of processing. The textile wet processing includes processes such as desizing, scouring, bleaching, dyeing, printing and finishing. Most of these processes employ hazardous chemicals which may lead to health hazards such as carcinogenicity, mutagenicity, respiratory problems, skin sensitisation/allergy, and environmental pollution. There comes the need for eco-testing of textile materials to ensure safety of the consumer.

10.2   Textile materials

Textile materials means the different fibre, yarn and fabric materials used in the manufacture of apparel, home textiles, technical textiles, etc. These materials are of different types and find various end uses. Classification of these materials is complex and for the purpose of testing the eco-friendliness and fixing the safe limits for certain hazardous substances Oeko Tex Standard 100 classifies textiles into four product classes as shown in Table 10.1.

Table 10.1. Classification of textile products by Oeko Tex Standard 100.

Product class	Description	Examples
I	Textiles and textile toys for babies and small children up to the age of three	Underwear, romper suits, bed linen, bedding, soft toys, etc.
II	Textiles which, when used as intended, have a large part of their surface in direct contact with the skin	Underwear, bed linen, terry cloth items, shirts, blouses, etc.
III	Textiles which, when used as intended, have no or only a little part of their surface in direct contact with the skin	Jackets, coats, facing materials, etc.
IV	Furnishing materials for decorative purposes	Table linen and curtains, textile wall and floor coverings, etc.

This classification aids in fixing the safer limits for the use or presence of hazardous substances in the product, e.g., the product class I sets very stringent regulation and lower limits of hazardous substances since this includes products for babies and toddlers who are very sensitive and delicate. Product class II is also crucial because they are in direct contact with the skin; longer exposure may increase the impact of hazards. Product classes III and IV are comparatively liberal in setting limits since their skin contact and exposure to the consumer is less compared to product classes I and II.

10.3   Importance of eco-testing of textiles

Importance of eco-testing of textiles are as follows:
* Ensures safety of consumer
* Aids in eco-labelling of products
* Serves as an eco-passport for international trade of textile and apparel products
* Gives indication of the pollution level
* Helps in the eco-management of the textile industry
* Avoids operational and environmental risks
* Helps in overall sustainable development.

10.4   Eco-parameters in textile testing

Basically, testing of textiles is done at various stages of the supply chain such as fibre, yarn, fabric and garment. Various mechanical properties like strength, dimensions, handle; and chemical properties like colour fastness, dimensional stability, skewness are tested based on the end use and buyer requirements. Apart from these basic tests, eco-testing have become essential for international trade since many countries restricts or regulates hazardous substances through government legislation and import of materials or finished products into the country need to comply with the legislation. Europe, especially the countries of the European Union (EU), has been the pioneer in environmental awareness. Regulations by European countries were the foremost that focused on consumer safety and environmental protection.

In case of textiles, eco-testing is a broad term, the reason being the use of enormous number of materials used at various stages of the production cycle. Hence, the parameters to be tested need to be communicated to the manufacturers. Table 10.2 shows the eco-test parameters and the standard test method adopted for testing. Based on the nature of the hazard, the use of certain substances is either completely banned or safe limit is fixed. There are two sources for fixing the safe limits for these eco-parameters: the government legislation or the brand's (buyer) requirement. At times, some of these values vary from country to country or brand to brand across the globe.

Table 10.2. Eco-parameters and test standards in textile testing

S. No.	Test Parameter	Permissible limits	Test Equipment	Testing Standard
1	Banned aromatic amines (present in azo dyes)	Germany: Total ban on azo dyes EU, India: < 30 ppm (mg/kg) China: < 20 ppm	Gas Chromatography – Mass Spectrometry (GC-MS); High Performance Thin Line Chromatography (HPTLC)	**EN 14362-1:2012** Determination of certain aromatic amines derived from azo colorants
2	Carcinogenic and allergic dyes	Prohibited from use Carcinogenic Dyes: 5 mg/l (as per law)	LC-MS LC-DAD	**DIN 54231: 2005** Detection of disperse dyestuffs **BVL B 82-02-10: 2007** Allergenic Disperse dyes and carcinogenic disperse dyes. Analysis of consumer goods. Verification of disperse dyes in textiles
3	Residual Pesticides	Babywear: 0.5 mg/kg Others: 1.0 mg/kg	GC-MS/GC-ECD/HPLC	Solvent extraction from fabric and detection on GC-MS.
4	Phthalates	< 0.1%	GC-MS	**ISO 14389:2014** Determination of the phthalate content — Tetrahydrofuran method
5	Chlorinated Phenols *Penta Chloro Phenol (PCP) and Tetra Chloro Phenol (TeCP)*	Children's wear: 0.05 mg/kg Adult wear: 0.5 mg/kg	HPLC/GC-MS/GC-ECD	**DIN EN ISO 17070:2007-01** Determination of tetrachlorophenol-, trichlorophenol-, dichlorophenol-, monochlorophenol-isomers and pentachlorophenol content

#	Parameter	Babywear (ppm or mg/kg)	Other textiles	Method	Standard
6	Formaldehyde	Babywear: 20 mg/kg; Items in direct contact with skin: 75 mg/kg; Items not in direct contact with skin: 300 mg/kg; Legal limit: 1500 mg/kg		UV-VIS spectrophotometer	EN ISO 14184-1 Determination of formaldehyde -- Part 1: Free and hydrolysed formaldehyde (water extraction method) EN ISO 14184-2 Determination of formaldehyde -- Part 2: Released formaldehyde (vapour absorption method)
7	Extractable (sweat/saliva) heavy metals			Thin Layer Chromatography (TLC) / Atomic Absorption Spectroscopy (AAS)	BS EN 16711-1:2015 Determination of metal content. Determination of metals using microwave digestion BS EN 16711-2:2015 Determination of metal content. Determination of metals extracted by acidic artificial perspiration solution
	Antimony (Sb)	30.0	30.0		
	Arsenic (As)	0.2	1.0		
	Cadmium (Cd)	0.1	0.1		
	Chromium (Cr)				
	- Textiles dyed with metal complex dyes	1.0	2.0		
	- All other textiles	0.5	1.0		
	Cobalt (Co)				
	- Textiles dyed with metal complex dyes	1.0	4.0		
	- All other textiles		1.0		
	Copper (Cu)	25.0	50.0		
	Lead (Pb)	0.2	1.0		
	Nickel (Ni)				
	- Textiles dyed with metal complex dyes	1.0	1.0		
	- All other textiles	0.5			
	Mercury (Hg)	0.02	0.02		

8	Nickel	Nickel release per week ≤ 0.5 µg/cm²/week	AAS Visual: using chemical	**BS EN 1811:2011+A1:2015** Reference test method for release of nickel from all post assemblies which are inserted into pierced parts of the human body and articles intended to come into direct and prolonged contact with the skin
9	Organotin compounds (DBT/TBT)	≤ 0.1 % DBT: 1.0 mg/kg TBT: 0.5 mg/kg	GC-MSD	**GB/T 20385 2006** Determination of organotin compounds
10	PVC (Poly Vinyl Chloride)	Prohibited from use	FTIR (qualitative analysis)	Bieltein test (burning method)
11	PFOA/PFOS (perfluorooctanoic acid/ perfluorooctane sulfonate)	EU: ≤ 1 µg/m² in coated textile material Others: 1 mg/kg	LC-MS or LC-tandem/ MS liquid chromatography / mass spectrometry	**EM201:2010** Test Method for the determination of PFOS and PFOA in Articles **PD CEN/TS 15968:2010** Determination of extractable perfluorooctanesulphonate (PFOS) in coated and impregnated solid articles, liquids and fire fighting foams. Method for sampling, extraction and analysis by LC-MS or LC-tandem/MS **ISO 25101:2009** Water quality -- Determination of perfluorooctanesulfonate (PFOS) and perfluorooctanoate (PFOA) -- Method for unfiltered samples using solid phase extraction and liquid chromatography/mass spectrometry
12	pH	between 4.0 and 7.5		**ISO 3071** Determination of pH of aqueous extract

No.	Parameter	Limit / Requirement	Method	Standard / Description
13	Flame retardants	< 5 mg/kg	GC-MS / LC-MS	ISO 17881-1:2016 Determination of certain flame retardants -- Part 1: Brominated flame retardants; ISO 17881-2:2016 Determination of certain flame retardants -- Part 2: Phosphorus flame retardants
14	PAHs (Polyaromatic hydrocarbons)	Benzo(a)pyrene: 1 ppm; Total of other 15 PAHs: 10 ppm	GC-MS	Solvent extraction from fabric and detection using GC-MS.
15	APEOs (Alkylphenol Ethoxylates)	EU Regulation: 1000 mg/kg; EU Ecolabel: Prohibited; Buyer requirement: 100 mg/kg	LC-MS	ISO 18254-1:2016 Method for the detection and determination of alkylphenol ethoxylates (APEO) -- Part 1: Method using HPLC-MS; ISO/NP 18254-2 Method for the detection and determination of alkylphenol ethoxylates (APEO) -- Part 2: Method using NPLC (This standard is under development)
16	Dimethyl formamide (DMF)	10 ppm or mg/kg	GC-MS	Extraction using organic solvent and detection with GC-MS.
17	Chlorinated organic carriers	< 1.0 ppm or mg/kg	GC-MS	DIN 54232: 2010 Determination of the content of bonds based on chlorobenzene and chlorotoluene
18	Acrylamide	REACH: restricted; Legal: 1000 ppm	GC-MS	Extraction with organic solvent and detection using GC-MS.
19	Glyoxal	Babywear: 20 ppm; Items in direct skin contact: 75 ppm; Outerwear: 300 ppm	GC-ECD	SN/T 4670-2016: Determination of glyoxal, formaldehyde and glutaraldehyde in textile fabrics - Liquid chromatographic method
20	Effluent parameters such as pH, TDS, TSS, TS, BOD, COD, DO, turbidity, colour, odour, etc.			

10.5   Test parameters and testing methods

Innumerable chemicals are used in the textile industry right from fibre production to the finishing of fabrics and garments. Some of the metal and plastic accessories used in clothing pose health hazards. Even the detergents used for laundering and the reagents used for dry-cleaning cause environmental pollution. Some of the chemicals and compounds identified to be very harmful, their nature, testing or analysis method are discussed the following sections.

10.5.1   Banned aromatic amines (present in azo dyes)

Azo dyes are popularly used in textile dyeing due to facts such as low-cost, brilliancy of colour obtained and good fastness characteristics. As the name indicates, it consists of azo group –N=N– and this azo group cleaves, under reductive conditions, to give two amines—NH_2 and some of these amines are found to be carcinogenic. There are thousands of azo dyes and pigments used and not all of them are harmful; only those azo dyes that release carcinogenic amines are banned or permitted within the allowed limits of 30 ppm (parts per million) or mg/kg. As per the REACH regulation and the Restricted Substances List (RSL), there are 24 such amines listed as carcinogenic and the Global Organic Textile Standards (GOTS) had listed 27 in total with three more amines added to the list.

To test the presence of these banned amines, the azo dye has to be extracted from the dyed textile material. The sample preparation technique is based on the fibre type: for fibres like cotton, viscose, wool and silk the azo dye is exposed to a reducing agent before extraction and then ethyl acetate is used to extract the dyes from the sample by solid phase extraction (SPE); for fibres like polyester, polyamide, polypropylene, acrylic or polyurethane the azo dye is extracted first and then the dye is reduced in the presence of sodium dithionite ($Na_2S_2O_4$) under mild conditions (pH: 7 and temperature 70°C). Then the amines are extracted by liquid–liquid extraction with t-butyl methyl ether (MTBE) and the resultant liquid is analysed by GC–MS (Gas Chromatography–Mass Spectrometry) and the results are expressed in parts per million (ppm) or in mg per kg.

10.5.2   Carcinogenic and allergic dyes

The carcinogenic dyes identified are listed in Table 10.3. They are prohibited from use.

Table 10.3. List of carcinogenic dyes.

Dye name (generic)	CAS number
C.I. Acid Red 26	3761-53-3
C.I. Basic Red 9	569-61-9
C.I. Basic Violet 14	632-99-5
C.I. Direct Black 38	1937-31-7
C.I. Direct Blue 6	2602-46-2
C.I. Direct Red 28	573-58-0
C.I. Disperse Blue 1	2475-45-8
C.I. Disperse Orange 11	82-28-0
C.I. Disperse Yellow 3	2832-40-8

Note: (Chemicals Abstract Service) - CAS number is a unique numerical identifier assigned by the Chemical Abstracts Service (CAS) to every chemical substance

Disperse dyes are a class of synthetic dyes that are water insoluble and used in dyeing of synthetic fibres like polyester, acetate and polyamide. Some of these disperse dyes, listed in Table 10.4 are identified to be causing allergic reactions and these are prohibited from use.

Table 10.4. List of allergic (disperse) dyes.

Dye name (generic)	CAS number
C.I. Disperse Blue 1	2475-45-8
C.I. Disperse Blue 3	2475-46-9
C.I. Disperse Blue 7	3179-90-6
C.I. Disperse Blue 26	-
C.I. Disperse Blue 35	12222-75-2
C.I. Disperse Blue 102	12222-97-8
C.I. Disperse Blue 106	12223-01-7
C.I. Disperse Blue 124	61915-51-7
C.I. Disperse Brown 1	23355-64-8
C.I. Disperse Orange 1	2581-69-3
C.I. Disperse Orange 3	730-40-5
C.I. Disperse Orange 37	-
C.I. Disperse Orange 76	-
C.I. Disperse Red 1	2875-52-8
C.I. Disperse Red 11	2872-48-2
C.I. Disperse Red 17	3179-89-3

C.I. Disperse Yellow 1	119-15-3
C.I. Disperse Yellow 3	2832-40-8
C.I. Disperse Yellow 9	6373-73-5
C.I. Disperse Yellow 39	-
C.I. Disperse Yellow 49	-
C.I. Disperse Orange 149	85136-74-9
C.I. Disperse Orange 23	6250-23-3

To determine the presence of these dyestuffs in textiles, the first step is to extract the dye from the fibre. Accelerated Solvent Extraction (ASE) or Speed Extractor is used for extraction at elevated temperature and pressure so that the resultant extract is almost colourless. The extract is then analysed by means of HPLC (High-Performance Liquid Chromatography) chromatogram.

10.5.3   Residual pesticides

Pesticides are widely used in cotton cultivation and wool production. Though the fibre undergoes many processes to be made into a garment, the garment contains residual pesticides. When exposed to direct skin contact, these pesticides have a tendency to enter the human body through the pores in the skin and also orally, in case of sucking of garments by children. Then the poisoning occurs with symptoms like headache, nausea, dizziness, vomiting, etc. Hence the use of pesticides is regulated in many countries. Certain chlorinated pesticides like Aldrin, Dieldrin, Chlordane, Endrin, Heptachlor, Hexachlorobenzene, Mirex, Toxaphene, Hexachlorocyclohexane are banned from use.

A study by Zameer, et.al, used soxhlet extraction method and ultrasound assisted extraction for extraction of residual pesticides from cotton fibre. Then the extract was analysed using Biosensor Toxicity Analyzer (BTA) by the measurement of bioelectrical signals caused by enzymatic inhibition of acetyl cholinesterase (AChE) for the detection of pesticides. Cryogenic homogenisation was carried out for sample pre-treatment.

10.5.4   Phthalates

Phthalates are group of chemicals used as plasticizers to improve the softness and flexibility of plastics, especially PVC. They are esters of ortho-phthalic acid and are used in making accessories, decorative items and plastic packaging. Phthalates are reported to be carcinogenic and endocrine disruptors in humans and animals. The use of phthalates is restricted in plasticised materials in toys and childcare articles and apparel with plasticised materials. Table 10.5 presents the list of restricted phthalates in textile articles and children's products.

BS EN 15777:2009 was the first method for determination of phthalate content in textile products. Later in 2014, a new standard ISO 14389 was issued and it supersedes the old standard BS EN 15777:2009. The old method employed soxhlet extraction with hexane as solvent while the new method employs ultrasonic extraction method with tetrahydrofuran (THF) solvent. The extraction time is 1 h and 4 h for the old and new methods, respectively. Then the extract is analysed using GC-MS and the phthalate content is calculated based on the ratio of the mass of the print or coating to the mass of the whole sample.

Table 10.5. List of restricted phthalates.

Name (phthalate)	CAS number
Di-(2-ethylhexyl) phthalate (DEHP)	117-81-7
Dibutyl phthalate (DBP)	84-74-2
Butyl benzyl phthalate (BBP)	85-67-2
Di-iso-nonyl phthalate (DINP)	28553-12-0 and 68515-48-0
Di-isodecylphthalate(DIDP)	26761-40-0 and 68515-49-1
Di-n-octyl phthalate (DNOP)	114-84-0
Di-isobutyl phthalate (DIBP)	84-69-5
Di-isohexyl phthalate(DIHP)	68515-50-4
Dipentyl phthalate (DPP)	131-18-0
Dimethoxyethyl phthalate (DMEP)	117-82-8

10.5.5   Chlorinated phenols

Chlorinated phenols are a group of substances with 1–5 chlorines covalently bonded to phenol and they include all isomers of mono-, di-, tri-, tetra-, and penta-chlorophenol. They are widely used as pesticides and preservatives for textiles. They are found in textile and leather materials, dyes and print pastes. Pentachlorophenol (PCP) and tetrachlorophenol (TeCP) are among the widely used chlorophenols and find application in making of apparel, footwear and accessories. Pentachlorophenol (CAS No. 87-86-5) is used as a preservative in textiles to protect from fungus, mould and insects. It finds major use in sizing starch preparation for warp beam. It is also used as a preservative in pigment emulsions, adhesives, glues, vegetable and mutton tallow. PCP is used as a biocide when textiles are to be stored or transported in humid conditions. Traces of PCP may be found in elastic rubber accessories and natural rubber latex based finishes. PCP released from products and dispersed into the environment can result in human exposure by breathing, eating, drinking, or absorbing the substance through the skin. Dispersion of substances that are persistent and that are stored in living organisms constitutes a special problem because the substances take a very long time to be reduced to a level that does not involve risk of damage. PCP is one of many health and environmentally hazardous substances.

Pentachlorophenol is very toxic, persistent and bio- accumulates in organisms. PCP is classified as "Toxic in humans in contact with skin and if swallowed" and as "Very toxic by inhalation". Damages have been recorded in the cardiovascular system, blood and liver when inhaled by humans. Tests on animals have shown that PCP has impacts on the immune system and central nervous system. It has also been classified as a carcinogen. Additionally, PCP is very toxic to many fish species. Dioxins and furans are formed as by-products in the production of PCP, which is the reason, products containing PCP are usually contaminated with dioxins and furans. PCP is an important source of emission of dioxins, furans and hexachlorobenzene. All these three substance groups have serious effects on health and the environment and are therefore strictly regulated.

The processes involved in the analysis of PCP are extraction, purification, separation, identification and quantification. ISO 17070 is a standard method for determination of PCP in leather and no standard method is available for textiles. A study by Su and Zhang (2011) devised a method for accurate measurement of PCP in textiles by isotope dilution liquid chromatography-mass spectrometry (LC-IDMS). Samples were pre-treated with acid and then with *n*-hexane and then measured based on isotope dilution LC-IDMS. The precision of this method is in the range of 0.80–1.40%. The method can trace to mass.

10.5.6   Formaldehyde

Formaldehyde is a volatile compound that is naturally present in small quantities. Human blood is said to have traces of formaldehyde. When formaldehyde is present in large quantities it may cause skin allergy or irritation, respiratory inflammation and eye irritation. The uses of formaldehyde in textiles include anti-shrinkage treatments, resin finishes for wrinkle or crease resistance and as dye fixing agents. Formaldehyde was first restricted in Japan, and then Finland regulated the use with limiting values. Now, many European countries have limited the use of formaldehyde. Two different test methods are performed to test the formaldehyde and are expressed as either free and hydrolysed formaldehyde or released formaldehyde. Free formaldehyde is a measure of the amount of formaldehyde present in the textile material while released formaldehyde is a measure of the amount of formaldehyde released by the material into the atmosphere. A common method employed for the determination of free formaldehyde in textile material is Pentane-2, 4-dione method which is also known as acetyl acetone method. The test procedure is as follows:

- Prepare standard formaldehyde solution by diluting 2.5 ml of standard solution (containing 1500 µg/ml formaldehyde) with water to 50 ml in a volumetric flask. This solution contains 75 µg/ml of formaldehyde.
- From the test sample prepare two test specimens that weigh 2.5 g each at an accuracy of 10 mg. Cut the test specimen into small pieces and transfer into a 250 ml flask with stopper.
- Add 100 ml of water to the flask and close tightly with the stopper. Place the flask in an ultrasonic bath at 40°C for 30 min. Then filter the solution and collect in another flask.
- Take 5 ml of the filtered solution in a test tube and 5 ml of standard formaldehyde solution in other test tubes. Add 5 ml of acetyl acetone reagent into each tube and shake well.
- Place the test tubes in a water bath at (40 ± 2)°C for (30 ± 5) min. Then keep at ambient temperature for (30 ± 5) min.
- Add 5 ml of acetyl acetone reagent solution to 5 ml of water and treat it the same way as the blank reagent. Measure the absorbance values in a 10 mm absorption cell at a wavelength of 412 nm against water in a UV-Vis Spectrophotometer and report the detected formaldehyde in mg/kg. If the concentration of formaldehyde is less than 20 mg/kg then the result is reported as "not-detectable".

The method used to test the released formaldehyde (during storage and ironing) is steam absorption or gas-phase method. The procedure to test the released formaldehyde by vapour absorption method is as follows:

- Take 1.0 g of test specimen from the textile sample and put it in a glass flask. Add water to the jar and seal it.
- Place the jar in an oven at 49°C for 20 h.
- Then the extract is analysed for the amount of formaldehyde vapour released by the material in water.

10.5.7   Extractable heavy metals

A heavy metal is a chemical element whose specific gravity is at least five times the specific gravity of water. Heavy metals are employed in various industrial applications such as the manufacture of pesticides, batteries, alloys, electroplated metal parts, textile dyes, steel, etc. It is manageable when smaller amount of these heavy metals are present in the environment while larger amounts may cause acute or chronic toxicity. Toxicity caused by heavy metals includes damage in central nervous function and damage of vital organs such as lungs, kidneys, liver. Long-term exposure results in neurological degenerative processes similar to Alzheimer's, Parkinson's disease, muscular dystrophy and multiple sclerosis. Allergy and carcinogenicity are also caused by repeated exposure to heavy metals.

Traces of heavy metals are often present in different textile process such as metal complex dyes, dye stripping agents, oxidizing compounds, antifungal, odour-preventive agents and mordant. The restricted heavy metals in textiles are Antimony, Arsenic, Barium, Cadmium, Chromium, Cobalt, Copper, Lead, Mercury, Nickel, Selenium, Tin, and Zinc. Heavy metals are the most commonly legislated chemicals globally.

Different analytical methods can be applied for the determination of heavy metals present in and on textile materials, in textile wastewaters, as well as in different reagent solutions used in textile processing. The process involves two basic steps:

1. Microwave digestion or extraction by acidic artificial perspiration solution/saliva solution.
2. Analysis.

The test procedure for extraction process is as follows:

- Dry the textile sample in an oven at $105 \pm 2°C$ at least for 4 h.
- Prepare the artificial perspiration solution as per the standard EN ISO 105-E04: In a 1000 ml glass flask, take 1000 ml of water and accurately weigh the following chemicals and add to the water and mix well.
- 0.5 g L-histidine monohydrochloride 1-hydrate
- 5.0 g sodium chloride
- 2.2 g sodium dihydrogenphosphate 2-hydrate.
- Adjust the pH of the solution to 5.5 ± 0.2 with dilute sodium hydroxide or dilute hydrochloric acid.
- From the test sample cut a specimen of 1 g and record the mass to the nearest 1 mg. If the test sample is heterogeneous, different parts are included as a composite specimen with equal parts of the material.
- Take the test specimen in a 100 ml flask and add 50 ml of the prepared artificial perspiration solution and shake by hand to ensure complete wetting.
- Then the specimen is shaken for 1 hour at $(37 \pm 2)°C$ in a shaker. Set the shaking frequency to 60 cycles per minute if using a horizontal shaker or 30 cycles per minute if using a circular shaker. A magnetic stirrer may also be used as an alternative.
- After the specified time, filter the extract so that solid textile particles and fluff are removed.
- The filtered solution is analysed by Microwave Plasma-Atomic Emission Spectroscopy (MP-AES).

The test procedure for microwave digestion method is as follows:

- From the test sample, prepare a specimen of 1 g and digest with 6 ml of 65% nitric acid (HNO_3) and 2 ml of 30% hydrogen peroxide (H_2O_2) in a microwave digestion system and diluted to 10 ml with deionised water. The advantage of microwave digestion is that it is a closed system.
- Then the digested solution is analysed by Atomic Absorption Spectrometry (AAS) and the concentration of heavy metals is expressed in μg/g or mg/kg.

10.5.8   Nickel

Nickel is used in metal accessories such as zippers, rivets, belt buckles, buttons and fashion jewellery. It is a highly allergenic heavy metal. Presence of nickel may pierce the skin and cause skin allergy such as contact dermatitis and hence the release of nickel by metal accessories is regulated and restricted. Test for nickel is done by three different ways as follows:

a. Nickel release per day
b. Nickel release per week ($μg/cm^2$)
c. Nickel spot test.

The original standard EN 1811:2011 was found to have uncertainty in measurement and hence the standard was amended on 15 January 2015 as EN 1811:2011+A1: 2015. The test procedure is as follows:

- The material to be tested is first cleaned and degreased.
- Artificial sweat solution is prepared and the material is placed it in and the temperature maintained at 30°C for a week (168 h).
- Then the extract is analysed using AAS (Atomic Absorption Spectrometry) or ICP (Inductively Coupled Plasma) Spectroscopy.

In case of metal coated items, the release value should not exceed 0.5 $μg/cm^2$/week for a period of 2 years of normal use of the item. The method for simulating 2 years of normal wear is defined by the standard EN12472:2009.

Another method for nickel test uses the standard PD CR 12471:2002. Screening tests for nickel release from alloys and coatings in items that come into direct and prolonged contact with the skin. The property of the nickel ion that forms coloured complex when comes in contact with dimethylglyoxime or dithiooxamide, is made use of in testing the nickel. To simulate the influence of sweat when in direct contact with the skin the material is pre-treated with artificial sweat. This is a quick screening method which provides guidance in evaluation of nickel release. The procedure is as follows:

- 0.8 ± 0.05 g of dimethylglyoxime is taken and dissolved in ethanol to make a 100 ml solution.
- 0.5 ± 0.05 g of dithiooxamide is taken and dissolved in ethanol to make a 100 ml solution.
- 5.6 g of sodium acetate trihydrate is taken and 2.4 ml of glacial acetic acid is added and dissolved with water to make a 10 ml solution.
- Artificial sweat solution is prepared by taking 1 ± 0.02 g urea, 5 ± 0.1 g sodium chloride and 1.13 ± 0.02 g lactic acid in a 2 l beaker and 1000 ml of deionized water is added and stirred well. The pH of the solution is adjusted to 6.5 ± 0.2 by addition of ammonia solution with stirring.
- Pre-test for nickel release: Moisten a cotton-wool-tipped stick with one or two drops of dimethylglyoxime solution and one drop of ammonia solution. Check that there is no discolouration. Rub firmly the cotton wool tip for 15 s against the surface to be tested. View the stick against a white background. The appearance of a red colour, from light pink to strong cerise, indicates nickel release.
- Laboratory test: Place the test item on a dish and pre-heat it to approximately 50°C. Using a Pasteur pipette, transfer one drop of artificial sweat onto the surface to be tested. Dry the item in the laboratory oven until its surface is completely dry. The temperature shall be 50 ± 3°C. Drying time will be about 15 min. Perform the test for nickel release.
- Field test: Place the test item on a dish and using a Pasteur pipette or a drop-dispensing bottle put one drop of artificial sweat on the surface to be tested. Dry the surface completely with the hair dryer. If necessary, keep the item in place using any suitable means. Place a thermometer close to the test item. The temperature shall not exceed 50°C.
- Testing for nickel release: Allow the item to cool for approximately 5 min. Moisten a cotton-wool-tipped stick with one or two drops of dimethylglyoxime solution and one drop of ammonia solution. Check that there is no discolouration. Place the tip of the cotton-wool-tipped stick onto the area to be tested and rub gently for 5 s. View the stick against a white background. The appearance of a red colour, from light pink to strong cerise, indicates nickel release. Using test strips for the detection of nickel: Allow the item to cool and immediately put one drop of ammonia solution on the dried salts. Place the test strip onto the treated area for 5 s. View the strip against a white background. The appearance of a red colour, from light pink to strong cerise, indicates nickel release.

Positive result: Formation of a red colour, from light pink to strong cerise, with dimethylglyoxime indicates that the nickel release from a tested surface is likely to be greater than 0.5 $\mu g/cm^2$/week, and a positive result should be reported. Formation of a black–violet colour with dithiooxamide confirms a positive result obtained with dimethylglyoxime.

Negative result: No colour change in both or either of these tests, indicates an absence of nickel release, and a negative result should be reported. However, such a result should be interpreted cautiously since these tests are of short duration and the test conditions are not comparable with those in EN 1811. The apparent absence of nickel release should be confirmed by EN 1811 or EN 12472 (followed by EN 1811), as appropriate.

Uncertain result: Apart from nickel, colouration in these tests can occur from metals such as cobalt, copper and palladium and mask any red colouration from nickel. Where colours other than red are obtained, the result should be reported as uncertain. In such cases, it is advisable to apply the appropriate reference method, EN 1811 or EN 12472 (followed by EN 1811).

10.5.9   Organotin compounds (DBT/TBT)

Organotin compounds are used in the following applications.

- As a biocide in preservation of cotton and polyester textiles
- As biocide for odour protection of sports textiles
- As a stabiliser or catalyst in PVC, polyurethane and polyester foams
- In anti-microbial finishing of textiles
- Organotin stabilisers may be found in polysiloxane which is used as a softener for polyester fabrics.

Li et al. (2011) developed a gas chromatography-mass spectrometry (GC-MS) method to determine the dibutyltin (DBT), tributyltin (TBT) and triphenyltin (TPhT) in textile auxiliaries. In this method, the sample was first extracted with *n*-hexane in acetate buffer solution (pH 4.0) by ultrasonication if the sample is hydrophobic or by oscillation extraction if the sample is hydrophilic. Then the extract is derived with sodium tetraethyl borate in tetrahydrofuran. The derivative is determined by GC-MS. The detection limits (Limit of Detection) were from 0.003 mg/l to 0.005 mg/l. The average recoveries of these organotin compounds at the three spiked levels of 4.0, 10.0 and 40.0 mg/kg were 92.6%–108.0% with the relative standard deviations (RSDs) of 2.5%–10.2%. The method is simple and accurate for simultaneous analysis of the DBT, TBT and TPhT in textile auxiliaries.

Zhiyuan et al., devised a method for simultaneous determination of organotin compounds (MBT, DBT, TBT) in textiles and leather using GC-MS. The method involved liquid/liquid extraction of organotin compounds from textiles and leather, derivation with NaBEt4 and analysis of the ethyl organotin by GC/MS in scan mode and selective ion mode. The detection limit was 0.01 mg/kg for both and the linearity response of all organotin compounds were good and R2 $\geq$0.999. The recoveries were 85.4~96.2%, and the relative standard deviations (RSDs) were 2.33~5.63%. The method is quick, accurate and sufficient for routine analysis in textiles and leather.

10.5.10   Polyvinyl chloride (PVC)

Polyvinyl chloride (PVC) is a thermoplastic polymer produced by polymerisation of the monomer vinyl chloride. It could be made softer and more flexible by adding plasticisers. It is used in making of garment accessories such as zip pullers, badges, sequins and as surface coating, plastisol prints on textiles. It is also used in making of raincoats and packaging materials. The cost of PVC is cheaper than rubber and latex and hence widely used. PVC has many adverse effects on the environment: it is environmentally persistent; the manufacture and disposal results in highly toxic wastes like dioxins. Vinyl chloride, used in making PVC, is a highly toxic compound and also carcinogenic. Beilstein test is performed to confirm the presence of PVC in materials suspected to contain PVC. Then it is further analysed by Global Apparel, Footwear and Textile Initiative (GAFTI) analytical method using Fourier Transform Infrared Spectroscopy (FTIR).

i. Beilstein test procedure
  - A copper wire is heated to red-hot with the flame from a Bunsen burner.
  - The hot copper wire is brought in contact with the test specimen so that the plastic melts and adheres to the wire.
  - Then the wire is again placed in the flame and the colour of flame at ignition is observed.
  - The green flame confirms the presence of PVC. This is because the chlorine atoms react with the copper wire, under flame, to form copper chloride which is volatile and emits a characteristic green light. Then the result is further confirmed by FTIR.
  - If the flame is colourless then PVC is absent in the test specimen and hence no further confirmation is required.

Caution: This test can generate toxic fumes and good ventilation must be ensured!

ii. GAFTI analytical method
  - The plasticiser from the test sample is removed by Soxhlet extraction with diethyl ether for at least 4 h. If the sample is not suitable for direct

FTIR analysis then it is dissolved in 5 ml of tetrahydrofuran (THF). If necessary, filtration is done to remove any interfering substances.

- Two laboratory control samples, positive and negative, are prepared.
- Using FTIR collect the spectrum of the control samples as well as the test samples. Then match the spectrum of the sample with compound in hit list.
- The characteristics peaks and/or bands are identified in the spectrum.
- The presence of PVC is confirmed if the characteristic major and minor absorption bands are present. The limit of detection is 10%.
- Characteristic major and minor absorption bands of PVC from spectrum of PVC:

    1. Broad band between 600–700 cm^{-1}
    2. Strong peaks at 1250 cm^{-1} and 1330 cm^{-1}
    3. Strong peak at 1430 cm^{-1}
    4. Double peaks at 2900 cm^{-1}.

10.5.11   Perfluorooctanoic acid/perfluorooctane sulfonate (PFOA/PFOS)

PFOA—perfluorooctanoic acid/perfluorooctane sulfonate ($C_8HF_{15}O_2$) is a fluorinated organic chemical that is either produced synthetically or through the degradation or metabolism of other flourochemical products. In textiles, PFOA is used in oil and water repellent finishing. They are persistent on the environment, bio-accumulative and toxic Persistent, Bio-accumulative and Toxic (PBT).

PFOS are similar to PFOA and used in the manufacture of rainwear, work wear, upholstery and home textiles such as bed linens, curtains, etc. PFOS are used in soil, oil and water repellent finishing of textiles. They are classified as PBT chemical compounds. The materials to be analysed are shown in Table 10.6

Table 10.6. PFOA and PFOS.

Materials to be analysed	Formula	Abbreviation	CAS number
Perfluoro-*n*-octanesulfonic acid (1,1,2,2,3,3,4,4,5,5,6,6,7, 7,8,8,8-heptadecafluoro-*n*- octanesulfonice acid)	$CF_3(CF_2)7SO_3H$	PFOS	1763-23-1
Perfluoro-*n*-octanoic acid (pentadecafluoro-*n*-octanoic acid)	$CF_3(CF_2)_6COOH$	PFOA	335-67-1

- From the test sample a specimen of about 2 g is taken from the middle part and weighed accurately to 0.001 g.
- The test specimen is further cut into tiny pieces and placed in the thimble of soxhlet extractor. About 150 ml methanol is taken in the round-bottom flask of 250 ml capacity. Then the test specimen is extracted in methanol for 6 h by reflux.
- The extract is concentrated to about 5 ml in a vacuum rotary evaporator. Then the solution is transferred to a 10 ml volumetric flask and filled with methanol.
- Standard PFOS and PFOA solutions are prepared and diluted with methanol. This is the calibration curve solution. At least four such standard solutions are prepared for each test.
- The test solution is injected in the HPLC system; the calibration curve solution is also injected under same conditions. A calibration curve is obtained in the peak area with the quantitative ion used; it is then compared with the peak area obtained from the test solution to be quantified.
- The amount of PFOS and PFOA are calculated and reported if the results are 10 mg/g in case of amount per unit weight and 1.0 $\mu g/m^2$ in case of amount per surface area. If the results are less than these values then it is reported as "Not detected".

10.5.12   pH

The term pH refers to the negative logarithm of the effective hydrogen ion concentration in a aqueous solution which ranges from 0 to 14, where 0–6.9 expresses acidity, 7 as neutral and 7.1–14 expresses alkalinity. Clothing is very intimate to human skin, which is slightly acidic in nature. Hence very high or low pH may cause skin irritation and itchiness. Clothes that come in direct contact with the skin is paid more emphasis on pH requirements which must be weakly acidic or neutral. The pH of textiles gets altered by the use of strong chemicals in processes such as scouring, bleaching, after-treatment and final washing. It is necessary to test the pH of the final product and give suitable treatment so as to neutralise, e.g., when the pH is strongly alkaline then acetic acid is used to neutralise it.

Higher pH value may exhibit yellowing tendencies, create change of shade upon storage and during shipment, result into poor dye pick up, patchy dyeing and poor colourfastness. Very high and very low pH results in degradation of textiles during storage and use, it may result in poor softness and harsh feel.

10.5.12.1   *Determination of pH of textiles*

- Cut the laboratory sample fabric into small pieces of about 5 × 5mm.
- It is preferred to use gloves for hands to prevent contamination. Also, material handling is minimised as much as possible.
- Wash the electrode of pH meter with distilled water and wipe with a paper napkin. Then calibrate the pH meter with either buffer solutions of pH 4 and pH 10.
- Weigh a test sample of 2.00 ± 0.05 g using precise electronic balance. Take three test samples as required by the client or buyer.
- Take the test sample in a conical flask and add 100 ml of distilled water or potassium chloride (KCl) solution; similarly prepare other specimens also.
- Cover the flasks and place in a mechanical shaker for 2 h. Record the temperature of the extracting solution used. In the meantime calibrate the pH meter using buffer solutions.
- Decant the first extract into a beaker and immediately immerse the electrode to a depth of at least 10 mm and stir gently with a rod so that the pH value stabilises. Ignore this value and do not record the pH of the solution.
- Decant the second extract into another beaker and immediately immerse the electrode to a depth of at least 10 mm and stir gently with a rod so that the pH value stabilises. Record the pH of the solution.
- Decant the third extract into another beaker and immediately immerse the electrode to a depth of at least 10 mm and stir gently with a rod so that the pH value stabilises. Record the pH of the solution.
- If the difference between the two recorded values is greater than 0.2, repeat the procedure with other test samples. Calculate the mean value when two valid measurements have been obtained.

10.5.13   Flame retardants

Flame retardants (FR) are chemical substances used in the finishing of textile materials meant for protective clothing (work wear), furniture upholstery, car interior furnishings and home textiles such as curtains, bed linen, etc. Various groups of FR are available in the market and are classified based on the chemical nature and functional behaviour. Modified polyester that is flame retardant by itself is safer than FR finished polyester. Based on the function FRs may be classified as either permanent or semi-permanent. Semi-permanent flame retardants impose health risks to the user and hence their use is banned. Flame retardants have tendency to bio-accumulate in the environment. They may be repro toxic and affect the immune system in human. If the products containing flame retardants are burnt or incinerated, dioxins and furans may be formed. Table 10.7 below lists the flame retardants and their CAS number.

Table 10.7. Flame retardants.

Name	CAS number
Polybrominated biphenyls (PBB)	59536-65-1
Penta-bromodiphenyl ether (pentaBDE)	32534-81-9
Octa-bromodiphenyl ether (octaBDE)	32536-52-0
Deca-bromodiphenyl ether (decaBDE)	1163-19-5
Tris(2,3-dibromopropyl)phosphate (TRIS)	126-72-7
Tris(1-aziridinyl)phosphine oxide (TEPA)	545-55-1
Hexa bromocyclododecane (HBCDD)	25637-99-4

For testing the presence of flame retardants in textiles, the test method employed is ISO 17881 that consists of two parts: Part 1: Brominated flame retardants—Part 2: Phosphorus flame retardants. The part 1 test procedure is as follows:

- Stock standard solutions of 1000 µg/ml each are prepared with individual flame retardants and toluene solvent. Similarly internal standard solution is prepared.
- Standard solution of 10 µg/ml decachlorobiphenyl is prepared in toluene.
- An admixture working solution of 17 flame retardants is prepared and diluted to a series of suitable concentrations depending on the requirements. At least five dilutions of the calibration sets are selected to create calibration curves and GC-MS analysis.
- Test specimen is prepared from the sample by cutting into small pieces that weighs (1.0 ± 0.01) g.
- The pieces are taken in a vial and 20 ml of toluene is added to it and closed tightly. The vial is placed in an ultrasonic generator and extracted for 30 min at room temperature. The extract is filtered and transferred into 100 ml flask. To the residue in the vial, 10 ml of toluene is added and extracted again for 15 min at room temperature. Then the extract is filtered and merged with the first extract in the flask. The extract is evaporated to near dryness and 2 ml of internal standard solution is added to dissolve the residue and then filtered.

- The filtrate is analysed for the flame retardants by GC-MS and the concentration of each flame retardant is quantified using the calibration curve. The results are expressed in µg/g.

10.5.14   PAHs

Polycyclic aromatic hydrocarbons (PAH) are a class of organic compounds that contain fused rings of carbon and hydrogen. They are used as raw material for the manufacture of dyes, rubber, plastics and pesticides. Based on their adverse effects they are classified as Persistent Organic Pollutants (POPs). There are 16 such PAHs identified and listed by (EPA) is given in Table 10.8. Among these Benzo(a) pyrene is most hazardous and is classified as class II carcinogen, mutagen and repro toxin. PAH may be found as contaminants in the lubricants used in textile processing. Carbon black based pigments or dyes may contain high concentrations of PAHs as impurities. Thermal decomposition of recycled materials may release PAHs during reprocessing. Long-term exposure to some PAHs may result in the development of particular cancers.

Table 10.8. List of PAH classified as POPs.

Name (PAH)	CAS number
Acenaphthene	83-32-9
Acenaphthylene	208-96-8
Anthracene	120-12-7
Benzo(a)anthracene	56-55-3
Benzo(a)pyrene	50-32-8
Benzo(b)fluoranthene	205-99-2
Benzo(ghi)perylene	191-24-2
Benzo(k)fluoranthene	2017-08-9
Chrysene	218-01-9
Dibenzo(a,h)anthracene	53-70-3
Fluoranthene	206-44-0
Fluorene	86-73-7
Indeno(1,2,3-cd)pyrene	193-39-5
Naphthalene	91-20-3
Phenanthrene	85-01-8
Pyrene	129-00-0

Li et al., devised a method for simultaneous determination of 16 polycyclic aromatic hydrocarbons (PAH) in textiles by gas chromatography-mass spectrometry (GC-MS). This method under optimised conditions exhibited good linearity with linear correlation coefficients between 0.9930 and 0.9999, with maximum recoveries within 74.0~99.0%, and with relative standard deviation (RSD) of 0.89–4.94%. Quantitation and detection limits ranged from 1.3 to 12.2 µg/l and 0.4–3.8 µg/l, respectively. These values were less than the limits of two relative standards (Oeko-Tex Standard 100 and ZEK 01.2–08). When applied to coated denim fabric and waterproof tarpaulins, results showed that PAH concentrations in waterproof tarpaulins were greater than the limits of the two relative standards.

10.5.15   Alkyl phenol ethoxylates (APEOs)

Alkyl phenol ethoxylates (APEOs) are non-ionic surfactants used in textile industry as detergents, industrial cleaners, dispersants, wetting agents, etc. They are also used in leather industry. Alkyl phenol ethoxylates are found to be hormone disruptors and toxic to aquatic organisms. Table 10.9 below lists the banned APEOs. Nonyl phenol (NP) and Nonyl phenol Ethoxylates (NPEOs) are the commonly used APEOs. Alkyl phenol ethoxylates lose their ethoxylates on degradation and form NP or OP those are extremely harmful.

Table 10.9. List of restricted APEOs.

Substance	CAS number
Nonyl phenol (NP)	25154-52-3
Nonyl phenol ethoxylates (NPEO)	9016-45-9
Octylphenol (OP)	27193-26-8
Octylphenol ethoxylate (OPEO)	9063-89-2

Nonylphenol ethoxylates belong to the non-ionic surfactant category and are of particular concern. The biodegradation of nonylphenol ethoxylate releases the branched nonylphenol, which is difficult to biodegrade. Nonylphenol is a substance having endocrine disruptive properties that can have serious effects on aquatic and many other organisms.

The standard ISO 18254 describes detection and quantification of extractable alkylphenol ethoxylates (nonylphenol ethoxylates and octylphenol ethoxylates) in textile products by liquid chromatography (LC)-mass spectrometry (MS) system. The textile sample is cut into small pieces, transferred to a vial, and extracted with methanol using ultrasound. The extract is filtered and not subjected to any additional cleaning. Subsequently, the methanol extract is analysed by LC-MS.

10.5.16   Dimethyl formamide (DMF)

N,N-dimethyl formamide is a polar aprotic solvent, miscible with water and majority of organic fluids. DMF is used in production of acrylic fibres and polyurethane products. It is also used in the manufacture of artificial leather, films and coatings. DMF is harmful if inhaled and while in direct contact with the skin. It is classified as a carcinogenic, mutagenic and repro toxic (CMR) substance and is being included in the candidate list of substances of very high concern (SVHC).

10.5.17   Chlorinated organic carriers

Chlorinated organic carriers are a class of compounds comprising of chlorotoluenes and chlorobenzenes with different substitution patterns of chlorine substitution. These compounds are used as carriers in disperse dyeing of synthetic fibres such as polyester acetates, polyacrylic polyamides, etc. These substances are toxic and found to affect the nervous system and may cause skin irritation. They may induce liver malfunction and irritation to mucus membrane. Hexachlorobenzene is classified as carcinogenic group 2 and 1,4-dichlorobenzene is classified as carcinogenic group 3. Table 10.10 lists the chlorinated organic carriers used in textile dyeing.

Table 10.10. List of chlorinated organic carriers.

Compounds	CAS number
2-Chlorotoluene	95-49-8
3-Chlorotoluene	108-41-8
4-Chlorotoluene	106-43-4
2,3-Dichlorotoluene	32768-54-0
2,4-Dichlorotoluene	95-73-8
2,5-Dichlorotoluene	19398-61-9
2,6-Dichlorotoluene	118-69-4
3,4-Dichlorotoluene	95-75-0
2,3,6 -Trichlorotoluene	2077-46-5
2,4,5 -Trichlorotoluene	6639-30-1
Pentachlorotoluene[	877-11-2
1,2-Dichlorobenzene	95-50-1
1,3-Dichlorobenzene	541-73-1
1,4-Dichlorobenzene	106-46-7
1,2,3-Trichlorobenzene	87-61-6
1,2,4-Trichlorobenzene	120-82-1

1,3,5-Trichlorobenzene	108-70-3
1,2,3,4-Tetrachlorobenzene	634-66-2
1,2,3,5-Tetrachlorobenzene	634-90-2
1,2,4,5-Tetrachlorobenzene	95-94-3
Pentachlorobenzene	608-93-5
Hexachlorobenzene	118-74-1

The method DIN 54232 employs extraction of chlorinated organic carriers from textiles by methylene chloride and detected with GC-MS.

10.5.18   Acrylamide

Polyacrylamide is used as additive in preparation of sizing starch and as stiffening agent in finishes such as water repellent and shrink proof finish. These textile materials may contain acrylamide monomer as residue. Acrylamide is stable under normal conditions, but may decompose or polymerise when heated. It is classified as category 2 carcinogen and mutagen. The European Chemical Agency (ECHA) has added acrylamide to the SVHC list. The fabric is taken from the sample and extracted with solvent and analysed with GC-MS. Accessories are also analysed for the presence of acrylamide.

10.5.19   Glyoxal

Glyoxal is a chemical used as intermediary in the dyestuff production. It is also used as cross-linking agent for polymer production, as a biocide and as a disinfectant. It is either used as direct cross-linking agent or as a building block for cross-linking agents that are used in textile finishes such as anti-crease, wrinkle recovery finishes. Glyoxal is used in the manufacture of reactant resins in textile industry. It is classified as category 3 mutagen and may lead to skin sensitisation and eye irritation. The test method for glyoxal involves solvent extraction and detection with GC-ECD at a detection limit of 10 mg/kg.

10.5.20   Effluent parameters

Effluent or the waste water from the textile industry possesses both organic and inorganic chemicals in high concentrations. The waste water can be discharged into water bodies only after the effluent treatment process so that the water is safe for human use and also for the aquatic life and environment.

The effluent from each process is distinctive and hence a common treatment may not be adequate. Hence the effluent parameters need to be analysed and treated accordingly. The important physio-chemical parameters to be tested in the effluent are as follows (Table 10.11):

- Chemical Oxygen Demand (COD)
- Biological Oxygen Demand (BOD)
- Total Solid (TS)
- Total Dissolved Solid (TDS)
- Total Suspended Solid (TSS)
- Dissolved Oxygen (DO)
- pH
- Acidity
- Alkalinity
- Colour and odour
- Heavy metals
- Nitrates
- Phosphates
- Turbidity
- Chlorides
- Sulphides.

Table 10.11. Eco-parameters in textile effluent.

Parameter (mg/l unless otherwise noted)	Limits
Temperature [°C]	15–35
TSS	50
COD	150
Colour [m⁻¹] 436 nm 525 nm 620 nm	7 5 3
pH	6–9
BOD	30
Total-N	20
Ammonium-N	10 l
Total-P	3
Oil and grease	10
Phenol	0.5
Coliform (bacteria/100 ml)	400
Cyanide	0.2
Sulphide	0.5

Sulphite	2
Persistent foam	Not visible
Heavy metals (mg/l)	
Antimony	0.1
Total chromium	0.2
Chromium VI	0.05
Cobalt	0.05
Copper	1
Nickel	0.2
Silver	0.1
Zinc	5.0
Arsenic	0.05
Cadmium	0.1
Lead	0.1
Mercury	0.01
Harmful substances (µg/l)	
AP and APEOs (all isomers)	5
Chlorobenzenes and chlorotoluenes	0.2
Chlorophenols	0.5
Dyes—Azo (forming restricted amines)	0.1
Dyes—Carcinogenic or equivalent concern	500
Dyes—Disperse (sensitising)	50
Flame retardants	5
Halogenated solvents	1
Organotin compounds	0.01
Perfluorinated and polyfluorinated chemicals *PFOS, PFOD, PFBS,PFHxA 8:2 FTOH, 6:2 FTOH*	0.01 1
Glycols	50
Ortho-phthalates—Including all ortho esters of phthalic acid	10
Polycyclic aromatic hydrocarbons (PAHs)	1
Volatile organic compounds (VOCs)	1

Apart from the above mentioned basic parameters, it is also important to ensure the absence of hazardous substances that are as follows:

1. Persistent, bio-accumulative and toxic (PBT)
2. Very persistent and very bio-accumulative (vPvB)
3. Carcinogenic, mutagenic and repro toxic (CMR)

4. Endocrine disruptors (ED).

10.6   Regulatory aspects

Textile industry persists globally and is a major economic contributor. The employment opportunities are numerous in the textile industry. The textile and apparel supply chain is long and involves many complicated production processes. Now, the making of textiles has become dangerous which urged the government to manage the textile industry with specific laws and regulations. The laws and regulations vary from country to country across the world. Table 10.12 list some of the regulations followed in some parts of the world.

Table 10.12. Textile regulation by different countries.

Country	Regulation
European union (EU)	• REACH Regulation No 1907/2006 Annex XVII • Textile Regulation (EU) No 1007/2011 • Short-Chain Chlorinated Paraffins (SCCP) Regulation (EC) No. 850/2004 and Regulation (EU) No. 519/2012 • Substances of Very High Concern (SVHC) in candidate list
Canada	• CCPSA, Restriction of Lead on Surface Coating Material in Children's Product (SOR/2005-109) • CCPSA, Restriction of Lead Content in Children's Product (SOR/2010-273) • CCPSA, Restriction of Phthalates in Children's Product (SOR/2010-298)
USA	• Consumer Product Safety Act (CPSA) • Consumer Product Safety Improvement Act (CPSIA) • Federal Hazardous Substances Act (FHSA) • Toxic Substances Control Act (TSCA)
Japan	Act on Control of Household Products Containing Harmful Substances (Act No. 112 of October 12, 1973)
India	• Textiles (Consumer Protection) Regulation 1988 • Environment (Protection) Act, 1986: Prohibition of 112 azo- and benzidine-based dyes
China	• GB 18401 National General Safety Technical Code for Textile Products • GB 31701 The Safety Technical Code for Infants and Children Textile Products • GB 5296.4 Instructions for Use of Products of Consumer Interest—Part 4: Textiles and Apparel • GB 20400 Leather and Fur—Limit of Harmful Matter • GB 21550 The Restriction of Hazardous Materials in Polyvinyl Chloride Artificial Leather

10.7 . Conclusion

A clear and profound understanding of the hazards caused by some of the dyes and auxiliaries used in textile production, especially the wet processing sector, paved the way to regulations by governments of countries across the globe and introduction of eco-label, both by the government and the industry. Now brands have become conscious about the need for eco-label and eco-friendly products. Thus, the testing of textile products for the presence of various banned or regulated substance is accomplished by eco-testing.

10.8   References

1. Zameer, Ul, Hassan, S., Jiri Militky, and Jan, Krejci, (2013), A Qualitative Study of Residual Pesticides on Cotton Fibers, Conference Papers in Science. Vol. 2013, Hindawi Publishing Corporation. 1–5

2. Islam, A. and Arun, K.G., (2013), Removal of pH, TDS and color from textile effluent by using coagulants and aquatic/non aquatic plants as adsorbents, Resources and Environment, 3(5): 101–114.

3. Su, F. and Zhang, P., (2011), Accurate analysis of trace pentachlorophenol in textiles by isotope dilution liquid chromatography - mass spectrometry, J Sep Sci, 34(5): 495–9.

4. Li, Y. et al., (2011), Determination of organotin compounds in textile auxiliaries by gas chromatography - mass spectrometry, Se pu = Chinese Journal of Chromatography, 29(4): 353–357.

5. Zhiyuan, W. et al., (2008), Determination of organotin compounds in textiles and leather by GC/MS [J], China Leather 13: 011.

6. Li, X. et al., (2016), Determination of polycyclic aromatic hydrocarbons in textiles by gas chromatography—Mass spectrometry, AATCC Journal of Research, 3(6): 6–11.

7. Gomathi, E., Rathika, G., and Santhini, E., (2017), Physico-chemical parameters of textile dyeing effluent and its impacts with case study, Int J Res Chem Environ, 7: 17–24.

8. Sivakumar, K.K. et al., (2011), Assessment studies on wastewater pollution by textile dyeing and bleaching industries at Karur, Tamil Nadu, Rasayan Journal of Chemistry, 4(2): 264–269.

9. https://www.blcchemicaltesting.com/chemical-testing/heavy-metals-testing-and-analysis/.

10. http://textilelibrary.blogspot.in/2009/03/disperse-dye.html.

11. http://www.pulcra-chemicals.com/zdhc.

12. http://www.centexbel.be/files/brochure-pdf/allergens_eng.pdf.

13. http://www.texanlab.com/documents/downloads/15.pdf.

14. http://www.dynamica-eu.com/literatures/TN2-Determination%20of%20 Formaldehyde% 20in%20Textile.pdf.
15. http://www.iisnl.com/pdf/iis17A05.pdf.
16. http://www.lookchem.com/C-I-Disperse-Blue-35/.
17. http://www.bpt.archroma.com/wp-content/uploads/2014/07/ZDHC-MRSL-Textile-Formulation. January 2018

11

Environmental sustainability in textile production and processing

Dr. P. SenthilKumar

Assistant Professor (Sr.Gr), Department of Textile Technology,
PSG College of Technology
Coimbatore-641 004

Environmental sustainability is a state in which the demands placed on the environment can be met without reducing its capacity to allow all people to live well, now and in the future. Furthermore, human well-being and the well-being of society is part and parcel of what sustainability stands for. Ensuring that human beings have access to basic resources, that their health is being protected, and that they enjoy a good quality of life within a sustainable environment is critical. The textile industry is considered as ecologically one of the most polluting industries in the world. The issues which make the life cycles of textiles and clothing unsustainable are the use of harmful chemicals, high consumption of water and energy, generation of large quantities of solid and gaseous wastes, huge fuel consumption for transportation to remote places where textile units are located, and use of non-biodegradable packaging materials.

11.1   Production and processing

11.1.1   Cotton cultivation

In cotton developing, lower the yield higher will be the natural effects. The impacts are centred around eco-toxicity, on the grounds that no other effect class shows a noteworthy effect. On account of development of natural cotton, the effect classes of eco-toxicity and summer brown haze are fundamentally influenced, trailed by the Green House Gases (GHG) impacts mostly caused by the compost connected and the gas utilised as the vitality source.[1]

Utilisation of reasonable crude materials, items, innovation, and vitality, e.g., crops utilising no pesticides or have a diminished requirement for water, utilisation of materials produced using sustainable assets with "elective green" substituted chemicals, and the utilisation of sustainable vitality are the accepted procedures at this stage. This can prompt monetary maintainability, as it was discovered that utilisation of natural over regular cotton decreases the number of nuisance administration days required every year by around 40%, thus the expenses of manures and irrigation administration falls essentially.[2]

11.1.2   Hazardous chemicals

The term "health hazard" incorporates chemicals which are cancer-causing agents or generally poisonous or profoundly dangerous specialists, which harm the lungs, skin, eyes, or mucous layers. A substance is assigned as a physical danger when there is deductively legitimate proof that it is a flammable fluid, a compacted gas, touchy, combustible, natural peroxide, oxidizer, pyrophoric, shaky (receptive), or on the other hand water receptive. Based on concoction conduct, in this way, perilous substances might be ordered as ignitable and combustible substances, oxidizers, receptive substances, or on the other hand destructive substances, yet maybe the best concern is with poisonous quality. Poisonous overwhelming metals and unpredictable natural mixes are two critical sub-gatherings of perilous substances. A few sorts of danger make it hard to characterise "safe" levels for substances, indeed, even at low measurements, for instance, substances might be:

- Carcinogenic (causing growth), mutagenic (ready to modify qualities), as well as repro toxic (hurtful to generation)
- Endocrine disruptors (meddling with hormone frameworks).

Overwhelming metals enter the earth through wastewaters from various branches of the material business, specifically from released turning showers, from man-made fibre producing plants, and from effluents released from colouring machines. The characteristics of green chemicals are as follows:[3]

- Prepared from sustainable or promptly accessible assets by naturally neighbourly forms
- Low inclination to experience sudden, fierce, capricious responses, for example, blasts
- Non-combustible or ineffectively combustible
- Low harmfulness and nonappearance of lethal constituents, especially substantial metals
- Biodegradable
- Low inclination to experience bio-collection in evolved ways of life in the earth. [4]

11.1.3   Man-made fibres—Polyester recycling

A comparison of the product life cycle from raw material to the point of sale was made with four different options:

- Recycling (by melt spinning)
- Reuse (by injection moulding)
- Household waste treatment (with heat recovery)
- Incineration with landfill.

Reusing, reuse, and burning all gave a 30% better positioning.

The outcomes from the individual procedure have just restricted significance for the entire life cycle of textiles all in all, on the grounds that there are numerous wandering parameters, especially regarding quality angles; besides, a similar texture quality created through various procedures on various hardware or with various formulas will bring about various impacts.

11.1.4   Yarn formation

11.1.4.1   Spinning and weaving

The fibre production process (cotton, polyester, or both) influences the environment more strongly than the fabric production method; the main impact categories in the production of open-end-spun (OE-spun) cotton yarns for denim fabrics are acidification and heavy metal contamination followed by winter smog (probably because of the chemicals used in fibre production/ cultivation).

A comparison of jeans fabrics produced with different spinning and knitting technologies shows that the highest impact on environment was with ring-spun jeans followed by OE-spun jeans; the lowest impact was with knitted shirts, even with equal weight as jeans. In fact, knitting technology has less impact on the environment than modern air jet weaving technology. However, advanced four-phase weaving was even more eco-friendly than conventional knitting because of its higher productivity. Fibre production influences the environment more strongly than the production processes. The environmental impact is greater in cotton growing than in polyester production. Hence, when fibre production is integrated with spinning and weaving, the impact is more with cotton than with polyester. [1]

11.2   Use of green technologies in the textile and clothing sector

11.2.1   Supercritical carbon dioxide dyeing

Supercritical carbon dioxide dyeing is water free process and therefore is a revolutionary and green attractive alternative to conventional wet methods in the textile industry. Supercritical carbon dioxide can be used as a medium in yarn preparation, colouration, and finishing of polyester, nylon, cotton, wool, silk, and other fabrics without generating aqueous effluents; this new method has many benefits, such as no wastewater, no auxiliaries, high dyeing rate, and good levelling results.

11.2.2   Electrochemical reduction of dyes

At present, a number of reducing agents like sodium sulphide, sodium hydrosulphite (sodium dithionite), hydroxyacetone, or glucose are required in the dyeing process for vat and sulphur dyes to ensure that these dyes stay in soluble leuco-form for adsorption and diffusion into the textile surface. The reduction mechanism of vat/indigo and sulphur dyes. However, because of the non -regenerable nature and formation of sulphite, sulphate, thiosulphate, and toxic sulphur, their use causes adverse health and environmental problems.

11.2.3   Microwave-assisted textile processing

Microwave technology involves electromagnetic radiation associated with electric and magnetic fields, which are oscillating at perpendicular directions with respect to each other. The use of microwaves as a heating source for desizing, scouring, and bleaching processes, dyeing, and drying processes has been well -documented in the literature. The application of microwave technology includes curing of cotton fabrics treated with no formaldehyde finishing agents such as glyoxal, glutaraldehyde and emerging green technologies and environment friendly products.

11.2.4   Plasma technology

In plasma treatment, the polymer/fibre surfaces are treated with excited and energetic plasma species (ions, radicals, electrons, and metastable). The plasma introduces new hydrophilic groups into the structure and has been used to achieve adhesion, or reflectivity, wettability, water repellency, dyeability, flame retardancy, and effective anti-microbial properties.

11.2.5   Ultrasound colouration

The application of ultrasound power is the most emerging and developing domain in the textile industry and has a big role in the concept of clean technology for textile processing. Ultrasound technology offers numerous benefits over the conventional chemical processes, mainly by generating cavitation in liquid medium in addition to other mechanical effects such as dispersion, degassing, diffusion, and intense agitation of liquid.

11.3   Fabric production

A woven fabric is the interlacement of warp and weft yarns. The warp yarns are applied with a suitable chemical —size, prior to the assembly on the loom, to give sufficient strength and protect against abrasion during the shedding process since the warp yarns are under stress and the weft yarns are inserted into the formed shed. This size makes the warp yarn stronger and more slippery by forming a film and hence reduces the number of yarn breakages.

11.3.1   Size

- Natural (made of potatoes, corn, maize, tapioca, etc.)
- Synthetic (carboxy methyl cellulose, poly vinyl alcohols, poly vinyl acrylates, etc.)

Cotton is the most heavily sized fibre with loads of 200 g/kg applied to the warp yarn. [5] The size preparation involves addition of other chemicals such as viscosity regulators, anti-static agents, wetting agents, defoaming agents and preservatives. Sizes for fibres other than cotton do not contain such a range of auxiliaries and generally only a preservative is added.

Desizing is the removal of the size paste. Man-made fibres are generally sized with water-soluble sizes easily removed by a hot water wash or in the scouring process. On the other hand, natural fibres such as cotton are most often sized with water insoluble starches or mixtures of starch and other size materials. Enzymes are used to break these starches into water-soluble sugars.

11.4   Wet processing

Scouring is a cleaning process which removes impurities from fibres, yarn or cloth. Scouring uses alkali to saponify natural oils, and surfactants to emulsify and suspend non- saponifiable impurities in the scouring bath. Bleaching is a chemical process which eliminates unwanted coloured matter from fibres, yarns or cloth. Bleaching decolourises coloured impurities not removed by scouring and prepares the cloth for further finishing processes such as dyeing or printing. The most common bleaching agents include hydrogen peroxide, chlorine bleaching, sodium hypochlorite, sodium chlorite and sulphur dioxide gas. [6]

Dyeing is the process of colouring the fabric or yarn or fibre. It is the major cause of water pollution, as it causes of colouring of the effluent. Pollutant impacts are also associated with chemicals used during dyeing, equipment maintenance and cleaning. Textile printing, similar to dyeing, also generates varying amounts and types of pollutants. Printing produces high BOD and COD loads only if preparation operations (scouring) are done on site. Print application consumes less water and produces lower BOD than preparation operations such as desizing, scouring and bleaching.

Most of these processes use water as a component and hence are termed as wet processing methods. The general approaches to textile waste reduction in textile wet processing are as follows:

- Preparation stage
- Recovery systems
- Waste steam reuse
- Chemical substitutions
- Alternative processing
  - Dyeing
- Reconstitution/reuse of dye-bath
- Chemical substitution
- Alternative processes
  - Finishing
  - Reuse
  - Substitution
- Alternative processes
  - General
- Waste characterisation
- Raw materials
- The fate of processing chemicals
  - Equalisation

11.5   Pollution in textile manufacturing and remedies

11.5.1   Air pollution

It is the most difficult to sample and test for study. Air emissions are classified based on the sources like boilers, ovens, storage tanks or diffusive ——solvent based, waste water treatment, warehouses and spills. Textile mills usually generate nitrogen and sulphur oxides from boilers. Other significant sources of air emissions in textile operations include resin finishing and drying operations, printing, dyeing, fabric preparation, and wastewater treatment plants. Hydrocarbons are emitted from drying ovens and from mineral oils in high- temperature drying/curing. These processes can emit formaldehyde, acids, softeners, and other volatile compounds.[7, 8] Acetic acid and formaldehyde are two major emissions of concern in textiles.

11.5.2   Water pollution

In the textile industry it is because of the usage of high volumes of water throughout its operations, from the washing of fibres to bleaching, dyeing and washing of finished products. The aquatic toxicity of textile industry wastewater varies considerably among production facilities. The sources of aquatic toxicity can include salt, surfactants, ionic metals and their metal complexes, toxic organic chemicals, biocides and toxic anions. Most textile dyes have low aquatic toxicity. On the other hand, surfactants and related compounds, such as detergents, emulsifiers and dispersants are used in almost every textile process and can be an important contributor to effluent aquatic toxicity, BOD and foaming. Wet processing creates the highest volume of wastewater of all the processing methods.[9]

The waste generated can be classified as reusable waste and non-reusable waste. This can be controlled or limited. The waste is usually a result of the following factors:

- Erroneous working methods
- Rejected materials, off specification materials, failed quality control
- Equipment malfunctions
- Poor housekeeping techniques.

11.6   Waste generation

Similarly as with some other industry, the textile industry produces all classifications of mechanical wastes, specifically fluids, solids, and gases. For greener procedures, non-renewable wastes should be reused and sustainable wastes should be treated the soil on the off chance that reusing isn't a choice. Different valuable materials can be recuperated from textile process wastes.

11.7   Conclusion

Every life cycle of a piece of clothing, starting from the raw material stage, followed by the manufacturing process, transportation and retailing, consumer use, and the disposal phase is responsible for the creation of various potential environmental threats. This is the crux of the problem that the textile industry is facing. It is imperative to make changes to the manufacturing and processing methods accordingly so that a sustainable environment is ensured. More emphasis on the formula Reject-Reduce-Reuse-Recycle should be given by all the stakeholders in the textile value chain. Environmentally safer alternatives for manufacture and processing should be used wherever possible. Thus it can be made sure that the next generations enjoy what the present generation people are enjoying now.

11.8   References

1. Tobler-Rohr, M.I., (2011), Handbook of sustainable textile production. Woodhead, Cambridge, 46–48.
2. Eyhorn, F., Maeder, P., and Ramakrishnan, M., (2005), The impact of organic cotton farming on the livelihoods of smallholders. Central India, 10–12.
3. Manahan, S.E., (2006), Green chemistry and the ten commandments of sustainability. Chem-Char Research Inc, Columbia, 10–12.
4. Roy Choudhury, A.K., (2013), Green chemistry and the textile industry. Text Prog, 45(1): 3143. doi:10.1080/00405167.2013.807601.
5. Karthik, T., and Gopalakrishnan, D., (2004), Environmental Analysis of Textile Value Chain: An Overview, in Road map to sustainable textiles and clothing, Springer, Singapore, 153–189.
6. Abdalah, G., Ramkumar, S.S., and Simonton, J., (1999), An investigation of environmentally friendly chemicals in Abdalah textile processes to reduce water, air, and odor emissions. Texas Food and Fibers Commission Annual Project Report, Retrieved 12 Sept 2004, from Texas Tech University Web site http://www.tffc.state. tx.us/TECHTXRL/PROGRESS/1998-1999/ environ.html.
7. Karthik, T. and Gopalakrishnan, D., (2012), Eco-friendly fibres of the future. Asian Text J, 21: 67–71.
8. Karthik, T. and Gopalakrishnan, D., (2012), Impact of textiles on environmental issues, Part-I, Asian Dyer, 9: 52–58.
9. Khan, T.I. and Jain, V., (1995), Effect of textile industry waste water on the growth and some biochemical parameters on Triticumaestivum. Journal of Environmental Pollution, 2: 47–50.